T0289179

Scientific Communication in History

Brian C. Vickery

The Scarecrow Press, Inc.
Lanham, Maryland, and London
2000

SCARECROW PRESS, INC.

Published in the United States of America
by Scarecrow Press, Inc.
4720 Boston Way, Lanham, Maryland 20706
http://www.scarecrowpress.com

4 Pleydell Gardens, Folkestone
Kent CT20 2DN, England

British Library Cataloguing in Publication Information Available

Library of Congress Cataloging-in-Publication Data

Vickery, B. C. (Brian Campbell)
Scientific communication in history / Brian C. Vickery.
p. cm.
Includes bibliographical references and index.
ISBN 0-8108-3598-3 (alk. paper)
1. Communication in science—History. I. Title.
Q223.V53 2000
501'4—dc21 99-058413

∞™ The paper used in this publication meets the minimum requirements of American National Standard for Information Sciences—Permanence of Paper for Printed Library Materials, ANSI/NISO Z39.48-1992.
Manufactured in the United States of America.

Contents

Illustrations

Tables

Preface

Science and technology are social activities that result in the cumulation and application of knowledge. Communication between scientists of ideas, methods and results is essential if cumulative growth of knowledge is to take place. Communication between scientists and technologists, and between technologists, is equally essential if knowledge is to be distributed to where it can be applied. Communication is thus an integral component of scientific and technological activity.

What motivates the individual scientist to enter into the communication process—to seek information, and to offer information? Many, no doubt, do so because they have internalized the logic of science—that its growth depends on cumulation, which can occur only if individual items of scientific knowledge are brought together in a meaningful way. Scientists who are synthesizers—like Aristotle or Ptolemy among the Greeks, or Gesner in the Renaissance, and many another compiler of encyclopedias or comprehensive handbooks—naturally spend much time seeking information from any and every source. Scientists who are researching a specific problem seek information about existing methods, data and theories that will facilitate their tasks. Scientists who have produced a new product or theory or method may seek within the sphere of practice for some activity to which the new knowledge may be fruitfully applied.

Information about the results that a scientist has achieved is communicated to the world for a variety of reasons. This information is the scientist's product, his intellectual property. He publishes it to gain some benefit from the work that he has put in—financial benefit, or prestige within the scientific community. In the sixteenth century, new results (particularly in mathematics) were announced in the form of a public challenge to rivals to try to solve the problem. Only in the eighteenth century did the scientific journal as a mechanism for announcing and claiming the credit for a new result fully come into its own. In some cases, publication hopes to elicit responses from other scientists, to bring about the formation of a new field of research or application.

Scientists and technologists transfer information to each other by meeting and discussion in institutions of all kinds—universities, academies and societies, establishments undertaking research or manufacture—as well as in periodic conferences. Face-to-face discussion has always been backed up by written correspondence (and most recently, by electronic mail). The transfer of information through space and time has been facilitated by graphic records of work done, at first manuscript, then printed, now increasingly electronic. Forms of record have included lecture notes, books, journal articles, technical reports and many more-specialized kinds.

Secondary guides to recorded information have been produced: encyclopedias, histories, bibliographies, catalogues and indexes, serial bibliographies such as abstracts journals and reviews of progress, compendia containing summarized and organized data. At times, translation of scientific and technical works from one language to another has been a major activity. Archives of scientific records have been established in libraries, collecting clay tablets, papyri and other manuscripts, as well as printed text, drawings, photographs and microcopies, and now electronic text.

This book aims to provide a brief historical overview of this communication process, paying particular attention to the origins of each form of communication. It relies almost wholly on secondary (sometimes even tertiary) sources, and its sole claim to merit is that it pulls together a great deal of diverse information. The choice of topics—and still more, the choice of examples to illustrate topics—is necessarily very selective, and for the modern period all too often with a British or English-language bias.

The word "scientific" in the title is used to embrace the fields of the natural sciences, technology, agriculture and medicine. However, the main characters in the story are not the scientists and technologists as such (the discoverers of new knowledge and the inventors of new techniques), but those who have helped to transmit and transfer knowledge among scientists and technologists—organizers of societies, publishers, compilers and bibliographers, editors and translators, librarians and information scientists, and those concerned with the terminology, nomenclature and classification of science. Some of these were, indeed, also discoverers or inventors, but it is their contributions to communication on which we will concentrate. If on occasion I refer to them as "men," the reader will understand that women contributed also to this work.

The book is addressed primarily to the growing community of those who are involved in facilitating the transfer of scientific and technical information, but it may be that others—interested either in science or in human communication in general—will find this account of use to them.

The work is the product of studies carried out at leisurely intervals throughout the course of a rather busy professional career as an information scientist. I have drawn in a few sections on material prepared by an erstwhile collaborator, Stanley Goddard. To whom else should I make acknowledgement? I have looked at so much material over the years that my intellectual and literary debt is limitless. Some small part of it is acknowledged below, and in the references at the end of the book.

> Communication is the essence of science.
>
> —Francis Crick, in a television program, 1977

> The scientific knowledge we possess is the result of a social endeavour, which over the centuries has developed an approach appropriate to its goals, and where the work of each individual is informed and controlled by that of his colleagues in this endeavour, of the past, the present and the future.
>
> —Jerome Ravetz, *Scientific Knowledge and Its Social Problems*, 1971

> Communism, in the non-technical sense of common ownership of goods, is an integral element of the scientific ethos. The substantive findings of science are a product of social collaboration and are assigned to the community. . . . The institutional conception of science as part of the public domain is linked with the imperative for communication of findings. Secrecy is the antithesis of this norm; full and open communication is its enactment. The pressure for diffusion of results is reinforced by the institutional goal of advancing the boundaries of knowledge and by the incentive of recognition, which is of course contingent upon publication. . . . The communal character of science is further reflected in the recognition by scientists of their dependence upon a cultural heritage to which they lay no differential claims. Isaac Newton's remark—"If I have seen farther it is by standing on the shoulders of giants"—expresses at once a sense of indebtedness to the common heritage and a recognition of the essentially cooperative and cumulative quality of scientific achievement.
>
> —Robert Merton, *Science and Democratic Social Structure*, 1942

> Frequent communication of ideas, and a regular method of keeping up such communication, are essential to works in which great labour and industry are

to be employed, and to which much time must necessarily be devoted; when the philosopher must not always sit quietly in his cabinet, but must examine nature with his own eyes, and be present in the workshop of the mechanic, or the laboratory of the chemist. . . . The mathematician, the astronomer, the mechanician, in the common intercourse of the world sees few men who have much sympathy with his pursuits. . . . The "world," to him, consists of a few individuals, by the censure or approbation of whom the public's opinion of him must be finally determined; with them it is material that he should have more frequent intercourse than could be obtained by casual encounter; and he feels that the society of men engaged in pursuits similar to his own, is a necessary stimulus to his exertions. Add to this, that institutions providing this opportunity for communication become centres in which information concerning facts is collected from all quarters.

—John Playfair, *Encyclopaedia Britannica*, 1816

It is true that, in science, more perhaps than in any other field of human enterprise, progress is possible, and has indeed largely occurred, without any knowledge of history; but such knowledge is bound to affect the future direction and course of science and, if the lessons of the past have been well read, progress will be quicker and surer.

—J.D. Bernal, *Science in History*, 1965

From the first writings onward a new sort of tradition, an enduring and immortal tradition, began in the minds of men. Life, through mankind, grew thereafter more and more distinctly conscious of itself and its world. It is a thin streak of intellectual growth we trace in history, at first in a world of tumultuous ignorance and forgetfulness. It is like a mere line of light coming through the chink of an opening door into a darkened room; but slowly it widens, it grows. At last came a time in the history of Europe when the door, at the push of the printer, began to open more rapidly. Knowledge flared up, and as it flared it ceased to be the privilege of a favoured minority. For us now that door swings wider, and the light behind grows brighter. Misty it is still, glowing through clouds of dust and reek. The door is not half open. Our world today is only in the beginning of knowledge.

—H.G. Wells, *Outline of History*, 1925

History enlarges the imagination, and suggests possibilities of action and feeling which would not have occurred to an uninstructed mind. Consider that today we see developing a type of man, endowed with all the hopefulness of the Renaissance or of the age of Pericles, persuaded that his more vigorous efforts can quickly achieve whatever has proved too difficult for the generations that preceded him. Ignorant and contemptuous of the aims that inspired those generations, unaware of the complex problems that they attempted to solve, his rapid success in comparatively simple achievements encourages his confident belief that the future belongs to him.

But to those who have grown up surrounded by monuments of men and deeds whose memory they cherish, there is a curious thinness about the thoughts and emotions that inspire this confidence; optimism seems to be sustained by a too exclusive pursuit of what can be easily achieved; and hopes are not transmuted into ideals by the habit of appraising current events by their relation to the history of the past. Whatever is different from the present is despised. That among those who contributed nothing to the dominion of Mammon great men lived, that wisdom may reside in those whose thoughts are not dominated by the machine, is incredible to this temper of mind. Action, Success, Change, are its watchwords; whether the action is noble, the success in a good cause, or the change an improvement in anything except wealth, are questions there is no time to ask.

Against this spirit, whereby all leisure, all care for the ends of life, are sacrificed to the struggle to be first in a worthless race, history and the habit of living with the past are the surest antidotes; and in our age, more than ever before, such antidotes are needed.

—Bertrand Russell, *On History*, 1904

I have dreamed that, despite the many errors inevitable in this undertaking, it may be of some use to those upon whom the passion for philosophy has laid the compulsion to try to see things whole, to pursue perspective, unity and understanding through history in time, as well as to seek them through science in space.

—W.H. Durant, *The Story of Civilization*, 1954

Introduction

Science in History

Science, it has been said, has two origins. On the one hand, it arises from systematizing, generalizing, and abstracting from practical observation and technique; on the other hand, from rationalizing and making concrete previous mythology and speculation. It lies between the lore of the craftsman or artisan and the traditional wisdom of the bard, shaman or priest. Both craft lore and priestly wisdom can be—and have been—transmitted by word of mouth, but the generalizations and explanations of science need to be recorded if they are to be effectively built into the cumulative store of scientific knowledge.

Who were and are the scientists? Science becomes distinguished from technology when practical knowledge of natural properties and craft procedures begins to be tied together by general propositions into systems of ideas. Specialized crafts have a long history: in ancient Mesopotamia there were physicians, smiths, dyers, tanners, weavers, fishermen, fowlers, potters, stonemasons and others with specialized knowledge and skills. But the specialization we call science for a long time did not establish any independent existence in society. "At the beginning," wrote Mumford, "chief, medicine man, magician, prophet, astronomer, priest were not separated functionaries."

In ancient times, a few of those in the more highly skilled crafts (such as doctors, builders, metal workers) might record observations and techniques, and generalize from them. But even when specific adepts in medicine, astrology or alchemy appeared, as well as architects and engineers, they were for many ages small groups funded by wealthy rulers, clerics, landowners or merchants. One might say that during this long period science existed and developed, but there was no clearly defined group that we can call scientists (the word itself was, in fact, not coined until 1840, by William Whewell). Communication in science during this period can then only be described in terms of the general methods whereby knowledge and information were transmitted in society at that time. And even at this level, there are large gaps in our knowledge: there is far too little evidence about the transmission of knowledge among the craft specialists, and between craftsmen and scholars.

Only in the last three or four centuries has science become established as a profession in its own right, with its specific education, literature and associations, so that it becomes possible to focus more clearly on methods of communication specific to science. Early in the modern period, many scientists were amateurs, living on family wealth or an income unconnected with science. But soon the typical occupation of the scientist became that of a teacher in a university, contributing to research in such time as could be spared from academic duties. By the time of the Renaissance, scientists were beginning to be interested in the work of the crafts, and interaction between the two streams of knowledge becomes more evident. In the late nineteenth and twentieth centuries research in part moved out of the university into industrial and governmental institutions, and technology and science became closely interwoven.

Technical and scientific records are first found in the ancient civilizations of Egypt and Mesopotamia (in this book we leave aside early developments in India, China and the Americas). From these sources, the main track of scientific advance is shown in figure 1. From its original sources, science spread by the sixth century B.C. to Greece; from there in the third century B.C. it was carried to Alexandria, and thence two centuries later to Rome and on to Constantinople (Byzantium). With the rise of Islam, Greek scientific knowledge was brought to Baghdad by the eighth century A.D. from Byzantium, and carried by the Arabs along the southern Mediterranean to Sicily and Spain. Meanwhile, a thin trickle of science spread from Rome north into Europe. In the eleventh century, Greek/Arab scientific knowledge was brought from Sicily and Spain into western Europe, later supplemented by a stream from Byzantium. From Europe, beginning in the eighteenth century, modern science spread throughout the world.

One view giving highlights of the development of science and technology is presented in figure 2, adapted from Bernal. It indicates the main historical eras and successive centers of scientific activity, key fields of scientific activity and technical development, and their interactions.

The Renaissance (1450–1600) was a decisive period in the history of science and technology. Printing provided for the first time an effective

Figure 1 The Track of Scientific Advance

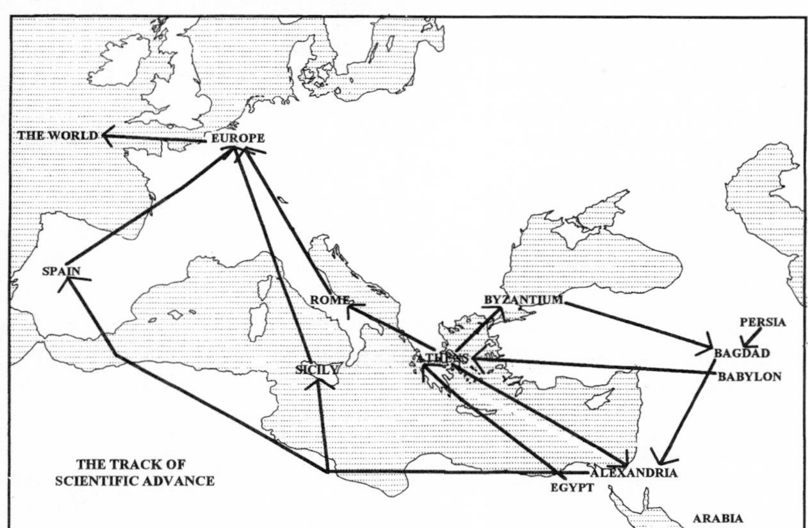

means of recording and disseminating new knowledge. The first independent forms of scientific association were established—academies and societies. The pace of scientific and technical activity greatly accelerated, and the growth of this activity has ever since been unabated. This growth is indicated in the next three figures. Figure 3 plots the growth in the number of printed books (in all subjects), based on studies by Iwinski and Barr. Figure 4, taken from Price, provides some indication of the development of scientific journals and abstracts journals over the period 1665–1950. Figure 5, from Vickery (1993), is an estimate of the growth in number of scientific and technical papers.

These publication growths reflect, of course, both the rising productivity of scientists and the spread of the habit of publication into applied science and technology. But above all they reflect a growth in the number of authors and of centers of scientific activity. With growth in the size of science has come ever-increasing specialization.

One further area of growth should be mentioned here—in the speed of communication. Transport is relevant to our story, because travel has always been important to scientists: to visit sources of information, to meet colleagues, to see scientific and technical products and techniques for themselves. The rate of travel on land was, until the nineteenth century,

Figure 2 Science and Technology in History

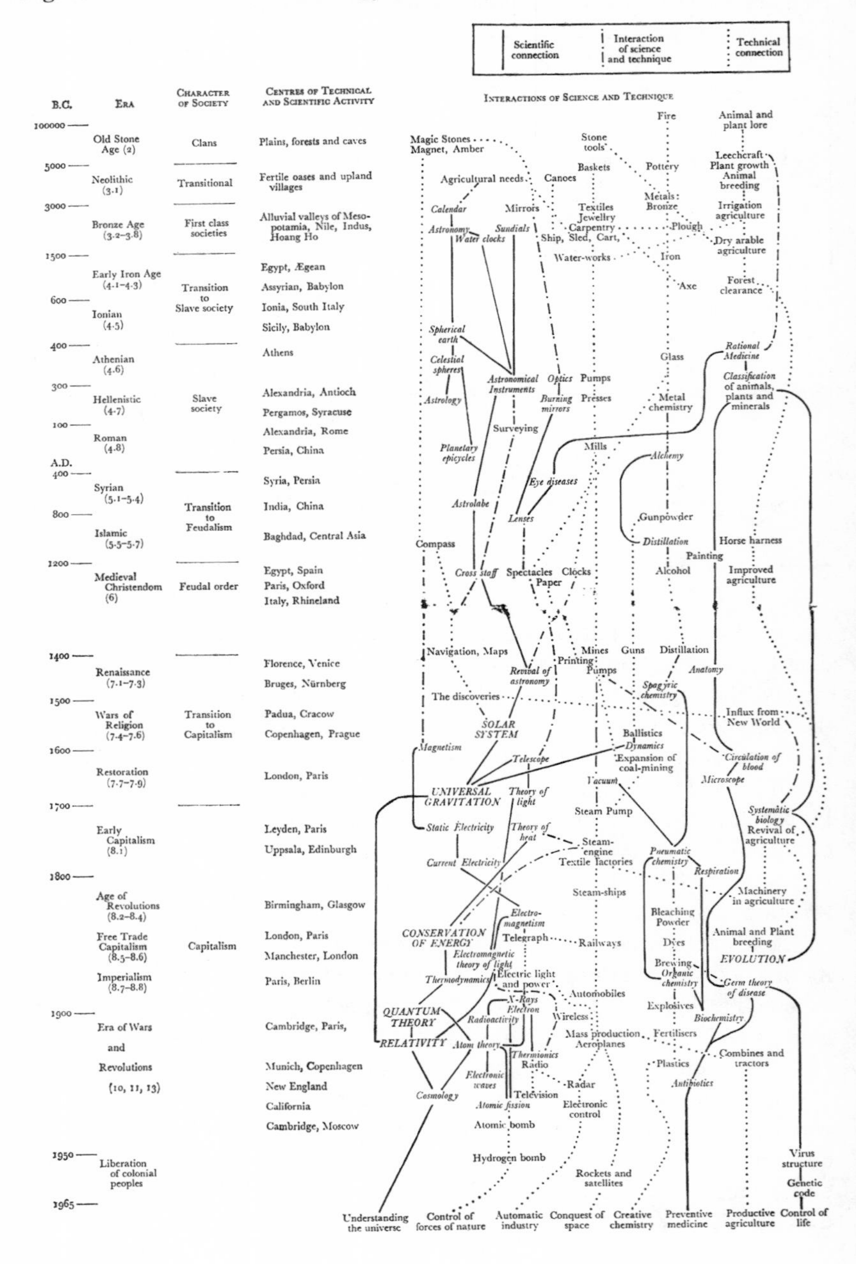

Figure 3 The Volume of Book Production (All Subjects)

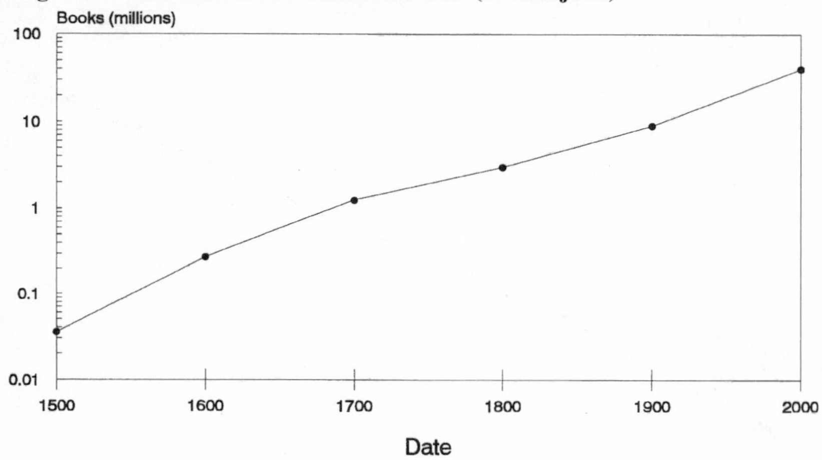

Figure 4 The Growth of Journals and Abstracts Journals

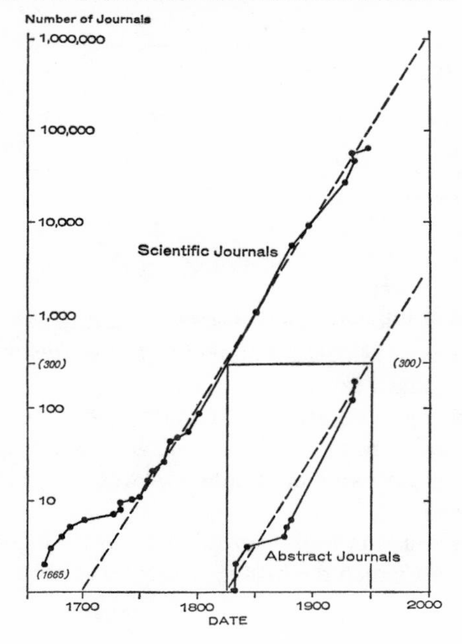

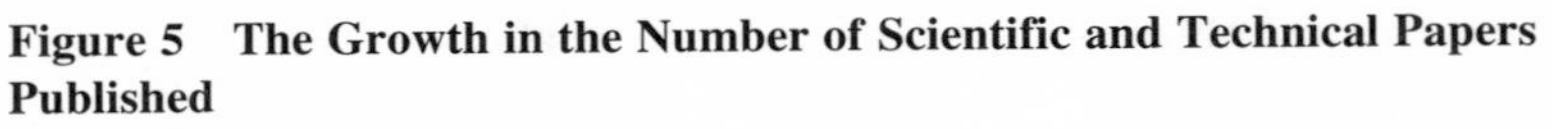

Figure 5 The Growth in the Number of Scientific and Technical Papers Published

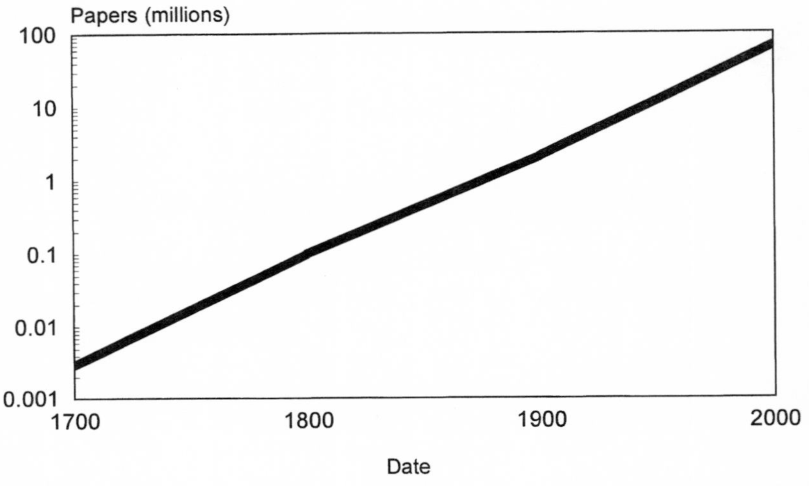

Table 1 Speed of Travel in Miles per Day

	3000 B.C.	*500 B.C.*	*1500 A.D.*	*1900 A.D.*	*Today*
Land	20	25	25	300–900	to 2,000
Sea/air	40	135	175	250	100,000

Source: McHale (1972)

limited to that of the horse-drawn coach (or, for urgent messages, the rider on horseback). The railway, the automobile, the steamship and the airplane changed that, as table 1 indicates.

Electronic telecommunications have made it possible for message transfer across the world to be almost instantaneous. Science is now worldwide, though as unevenly distributed among the world's peoples as any other cultural or material good.

This brief introduction to the spread and growth of science provides a background against which the history that follows may be set.

Section 1

The Earlier Civilizations (to about 600 B.C.)

THE RISE OF CIVILIZATION

There is little that we could truly characterize as scientific activity before the flowering of science in Greece, but long before this, man had amassed a great amount of practical knowledge about his physical and biological environments, and had developed many techniques to manipulate them. Urban civilization arose about 3000 B.C. around the Nile, Tigris-Euphrates and Indus rivers, but Childe has said of the 3000 years *before* this that "in no period of history till the days of Galileo was progress in knowledge so rapid or far-reaching discoveries so frequent."

Let us glance at man's technical equipment in 3000 B.C. He had already for millennia possessed axes, knives and scrapers of stone, spears and harpoons, needles of ivory, the bow and arrow, the bow-drill, fish hooks and nets. He had turned from gathering food to producing it—growing wheat, barley and millet; irrigating, plowing and manuring his fields; reaping his corn with a sickle, threshing it with a flail, grinding it in a quern. He cultivated many vegetables and fruits. He spun fibers on a spindle and wove them on a loom, making thread, rope, clothing and baskets from them. First domesticating the dog, he had gone on to keep sheep, cattle and pigs. He made and used fire for cooking and for operations on clay and metals. He fashioned clay vessels on a potter's wheel, baking and, later, glazing them. He built brick houses. He made wheeled carts and harnessed oxen and asses to them, he hoisted sails to the masts of his ships. He quarried for copper ore and smelted it, then hammered or cast the copper into a variety of utensils, tools and weapons.

Through the practice of tool making and tool using, men learned the mechanical and thermal properties of many natural products, thus laying the basis of physical and chemical science. Even at the food-gathering stage, man was building up a stock of plant and animal lore that would form the basis of biological science, and the use of herbs in illness looked forward to medical science.

The growth of cities after 3000 B.C. was a consequence of man's technical knowledge and of the increased productivity of his labor. Copper was replaced by the harder alloy bronze, and later by iron, and there were many improvements in the tools of the smith, the mason, the carpenter and the farmer. Glass was discovered, cast and molded into many shapes. After 2500 B.C., urbanization ushered in a cultural revolution: here arose the first known attempts to make a calendar, to study the movements of the stars, to investigate mensuration, to devise a system for writing numbers, and to observe and reflect on the anatomy of the human body and the injuries that it may sustain.

A close connection between urbanization and the written record has been described by Mumford:

> It is no accident that the emergence of the city as a self-contained unit, coincided with the development of the permanent record, with the first abstractions of number and verbal signs. By the time this happened, the amount of culture to be transmitted was beyond the capacity of a group to achieve orally. It was no longer sufficient that the funded experience of the community should repose in the minds of the most aged members.
>
> In daily transactions, the same need for permanent notations and signs was even more obvious: to act at a distance through agents and factors, to give commands and make contracts, some extrapersonal device was needed. The earliest tablets from Mesopotamia are mere lists and tallies—amounts of flour, bread, beer, livestock, etc, bare factual notations for enabling the community to keep track of quantities.
>
> The control of such activities was at first largely in the hands of a priestly class, freed from the constant necessity of manual labour, and increasingly conscious of the mediating function of mind. By progressive degrees of abstraction and symbolisation, they were able to turn the written record into a device for preserving and transmitting ideas and feelings and emotions that had never before taken any visible or material form.

MESOPOTAMIAN CULTURE

The phrase "ancient Mesopotamia" may be used to represent an historical epoch that stretched from before 3000 B.C. to about 500 B.C. Its source and heartland was "the land between the rivers" of Tigris and Euphrates, running through modern Iraq into the Persian Gulf, but the empires formed during this period extended at times to the Mediterranean coast and even to Egypt.

Villages had begun to develop into cities from 3000 B.C. Many different cities were at one time or another the center of early empires—such as Nippur, Ur, Lagash, Assur, Mari, Babylon, with populations of ten, twenty, or even thirty thousand. The last-mentioned city gave its name to the biblical Tower of Babel, the stepped temple characteristic of Mesopotamian cities since their early formation. The priests of the temples at first administered the cities, supervising granaries, storehouses, workshops and the work on the surrounding land. No city was self-sufficient: trade was essential to import metal, timber, stone, jewels and other wares. As the population grew, and with it pressure on the land, cities came into conflict with one another. Military leadership became important, and this developed into kingship. Inter-city warfare led to conquests and eventually to the growth of wide-ranging empires.

Within these empires there were problems of transport and communication. In the third millennium B.C., the chief land transport was by porters, oxen and asses—the horse was not introduced until about 1500 B.C. Speed was inhibited by the lack of good roads—twenty miles per day was the most that could be expected. Sailing ships were effective—and faster—up and down the rivers and even around the coasts. Travel was probably restricted to traders and government officials. A postal service for administrators seems to have been in existence.

One outstanding and unifying feature of Mesopotamian development was the physical nature of the written record: inscriptions on clay tablets. This prevailed during the whole period. Clay was used for writing not only in Mesopotamia, but also in neighboring territories to the east, north and west, and in the Mediterranean cultures of Knossos, Pylos, Mycenae and Thebes. For more than half the time that mankind has communicated in writing, much of the writing was on clay—a soft clay tablet was impressed with marks by a stylus, then dried in the sun or a kiln. Early pictorial logographs were soon replaced by symbols built from a set of wedged straight lines, cuneiform script.

This script was probably first used by Sumerians, a people living in the southern part of Mesopotamia, who spoke a language which appears not to be related to any other known, and which died out of common use by about 2000 B.C. By then the script had been adopted by the Akkadians, living further to the north, who spoke a Semitic language with two main dialects, Babylonian and Assyrian. Other linguistic groups later adopted and adapted the same cuneiform script.

In Mesopotamian clay tablet collections, eventually a variety of different kinds of text were to be found: administrative records; collections of laws (of which the most famous is the code of Hammurabi of Babylon, eighteenth century B.C., actually found inscribed on stone); governmental edicts; legal agreements; official letters; sacred myths; ritual texts; collections of proverbs; and some technical texts—astronomical ephemerides, chemical recipes, medical prescriptions, multiplication tables.

To produce this variety of text required the education and training of scribes, attached as skilled administrators to temples and courts. Students learned to construct clay tablets and styluses, and to inscribe cuneiform symbols. Word lists were used as teaching aids, some arranged in subject groups: trees, wooden objects, stars, clothes, animals, foods, parts of the body. As well as writing skills, mathematics and music were taught. Such scribal schools were attached to temples and palaces.

There is clear evidence of the existence of archival and library collections of tablets. Perhaps the earliest so far discovered is the collection of three hundred tablets found in the royal palace of Mari, dated to 2400 B.C., dealing mostly with the administration of the king's affairs. Also at Mari, a collection of twenty thousand tablets was uncovered, from about 1800 B.C., again mostly administrative. Indirect evidence exists for other long-continued collections of information—astronomical observations, leading to calendrical and astrological calculations. Scholarly libraries are less in evidence until later, but there is one collection of twenty thousand tablets made by Assurbanipal, king of Assyria, at Nineveh, dated about 650 B.C.

EGYPT, PERSIA AND PHOENICIA

The earliest records of Egypt depict a country already united under a single king. Writing in Egypt began soon after (and perhaps was influenced by) its appearance in Mesopotamia. The earliest texts are stone inscriptions with pictorial logographs. This hieroglyphic writing developed into a syllabic form, always retaining strongly pictorial features. About 2500 B.C. a more cursive form appeared, termed hieratic, and much later a further simplified demotic.

Relatively few early Egyptian texts have survived, because of the destructibility of the papyrus (sheets made from the leaves of the papyrus plant) on which they were written. Unlike clay tablets, a discarded papyrus document could be used as fuel, or be torn up and used for wrapping, or be used again for rough work. Its chance of survival was very small.

There is therefore little physical evidence of document collections, but much indirect evidence. As early as 2500 B.C., tomb inscriptions describe royal establishments such as "the house of the sealed writings" and "the house of the archives," and there is much reference to scribes, who were general administrators as well as being responsible for records. One papyrus sets out some of the attainments to which a scribe attached to the army should aspire: in addition to knowing the classic texts of Egypt, he should be able to calculate the distribution of rations to soldiers, to estimate the number of bricks needed for construction of a ramp, to plan and supervise the construction and erection of an obelisk or statue, to organize supplies for a military expedition.

Another early Egyptian papyrus, purporting to be a letter from a father to his son, illustrates the gulf between the scholarly scribe and the craftsman:

> I have considered manual labour—but give your heart to letters. . . . I have seen the blacksmith, directing his foundrymen, and I have seen the metal-worker at his toil before a blazing furnace. His fingers are like the hide of the crocodile, he stinks more than the eggs of fish. . . . And every carpenter who works or chisels, has he any more rest than a ploughman? . . . The weaver sitting in a closed-up hut has a lot that is worse than that of a woman. . . . Truly, the occupation of scribe is the best of all. . . . One day in the chamber of instruction is better than eternity outside it . . . there exists no scribe who does not eat the food of the king's house.

Apart from administrative records, papyri of other kinds have been found, and these include collections of arithmetical and geometrical problems, and of medical prescriptions and surgical practices. But all in all, it is difficult to build up much of a picture of Egyptian scientific activity and the transfer of technical information.

Let us turn to another area. The Iranian plateau lies to the east of Mesopotamia. The area closest to the "two rivers" was Elam, and here occurred the growth of cities such as Susa, and of a cuneiform script derived from Sumeria. Contact with Mesopotamia continued throughout the millennia, and from time to time Elam was under Babylonian or Assyrian rule. In the seventh century B.C., Assurbanipal's armies destroyed Susa, and Elam as a separate entity disappeared from history.

Some two centuries earlier, the first mention is found of the Persians, an Indo-European people who had entered the plateau from the northeast. By 550 B.C. their ruler Cyrus established control of the Assyrian empire, and his armies spread east and west—to the borders of India, into Egypt, to the Mediterranean. They at first adapted Elamite cuneiform for their own language, but within a few centuries this was replaced by a more cursive script, and later still, by Arabic script. Apart from this, the Persians also, in the fifth century B.C., adopted as the official language of their empire the Semitic language Aramaic, which had become the lingua franca of the whole region.

This Persian interlude is important to our story in several ways. First, the domination of Mesopotamia and its neighbors for several centuries by rulers with quite a different language, a different script, and a different cultural background must have done much to accelerate the decline and eventual disappearance of the ancient cuneiform. The old collections of Babylonian and Assyrian tablets became dust-covered, buried at last under desert sands. Second, we shall find the Persians later playing an important part in the transmission of Greek science.

A third aspect is the development of the physical means of communication. As already mentioned, the speed of land transport in Mesopotamia had been restricted by the lack of good roads. The so-called "royal road" of the Persians ran from Susa near the Persian Gulf to Sardis on the Aegean, sixteen hundred miles away. The level and partly paved tracks were policed by guards posted at resthouses every fifteen miles. Special messengers, with changes of horses, could attain a speed of one hundred miles a day along these lines of communication. Along these roads later traveled the conquering Alexander of Macedon, who ended the first Persian empire.

Our last note on civilization "before the Greeks" concerns the Phoenicians. Since they lived on the east shore of the Mediterranean, it is not surprising that they were active sea-goers. From their home cities—Byblos, Berytus (Beirut), Sidon and especially Tyre—they sailed to Egypt, Cyprus, Crete, the north African coast, Spain, and even perhaps to the British Isles. In all these places colonies were set up, of which Carthage in north Africa, founded in the ninth century B.C., became the most famous. The Phoenicians have long been credited with the development of alphabetic script, and during the twentieth century inscriptions have been found to confirm this. The earliest, dated about 1000 B.C., comes from Byblos (from which the word "bible" is derived). This is considered to be the prototype from which later Phoenician, Aramaic and eventually all other alphabetic scripts (Greek, Latin, Arabic, etc.) were derived. It should be mentioned that an earlier alphabetic script, with signs for letters made up of cuneiform wedges, has been discovered at Ugarit in Asia Minor and dated to 1500 B.C., but it was the Phoenician signs that were to carry the day.

Section 2

Classical Culture (600 B.C. to A.D. 500)

THE EMERGENCE OF GREECE

The group of languages we call Indo-European appears to be descended from a common tongue spoken by peoples living north of the Caspian and Black Seas. The peoples began to disperse around 2000 B.C., and their languages diverged, giving rise to Persian and Sanskrit, and to the Germanic, Slavic and Celtic languages. Some groups reached the Mediterranean, occupying central Italy, where Latin developed. Others filtered into what is now Greece during the period 1850–1600 B.C.

The earliest evidence of written records of the Greek language comes from Crete. A syllabic script dating from 1400 B.C. was discovered on clay tablets at Knossos in 1909 by Arthur Evans, who named it Minoan Linear B. In 1952 it was established that the language was Greek. Linear B texts were also found in mainland Greece, for example at Thebes and Mycenae. It was at this period that the Mycenaeans launched an expedition against their commercial rival, Troy in Asia Minor, an event later commemorated by Homer.

In about 1200 B.C., a new wave of Greek speakers, the Dorians, invaded the peninsula, and the use of the syllabary died out. The first visual evidence for an alphabetic Greek script appears on vase and rock inscriptions dated about 700 B.C., although it had probably been adopted somewhat earlier. It is likely that Homer's heroic epics, the *Iliad* and the *Odyssey*, were committed to alphabetic writing at that time, as well as works by the Greek poet Hesiod and the oral tradition of law.

Greek alphabetic script was derived from the earlier script developed by the Phoenicians. It seems that the simplification introduced by the alphabet took the mystery out of writing. Such scripts made possible the spread of literacy among wider sections of society, and writing was no longer the preserve of conventionally minded priests, scribes and rulers. Some evidence of instruction in literacy in the seventh century B.C. is provided by an "alphabetarion" of that date: an ivory tablet inscribed with a Phoenician-style alphabet, a Semitic alphabet, and some Greek letters. Tablets of this kind, coated with wax, were later often used to practice writing, using a stylus to copy the inscribed letters as models.

The democratization of learning implicit in simplified scripts had major cultural consequences. The Greeks moved towards intellectual speculation of a kind that had no precedent in the early civilizations. These philosophical speculations by individual thinkers formed the matrix from which science slowly began to emerge.

Paradoxically, Greek thinkers gave little importance to the art of writing. It became commonplace—a technical task that could perhaps be handed over to slaves. In *Phaedrus*, Plato put into the mouth of Socrates a fable that seems even to denigrate the art:

> I heard that in Egypt there was one of the ancient gods of that country, the one whose sacred bird is the ibis, and whose name is Thoth. He it was who invented numbers and arithmetic and geometry and astronomy and, most important of all, letters. Now Thoth came to the king of all Egypt to show his inventions, saying that they ought to be imparted to the other Egyptians. But the king asked what use there was in each. . . .
>
> When they came to the letters, Thoth said "This invention will make the Egyptians wiser and will improve their memories, for it is an elixir of memory and wisdom that I have discovered." But the king replied "One man has the ability to beget arts, but the ability to judge of their usefulness to their users belongs to another. You, the father of letters, have been led by your affection for them to ascribe to them a power the opposite of that which they really possess.
>
> "For this invention will produce forgetfulness in the minds of those who learn to use it, because they will not practice their memory. Their trust in writing, produced by external characters which are no part of themselves, will discourage the use of their memory within them. You have invented an elixir, not of memory, but of reminding; you offer your pupils the appearance of wisdom, not true wisdom, for they will read many things without instruction and will therefore seem to know many things, when they are for the most part ignorant."

In 750 B.C. there began a period of Greek colonization, during which city-states were founded both in mainland Greece and in many other places: in Ionia (Asia Minor), and in the areas we now know as Sicily, Italy, southern France, Egypt, and around the Black Sea. By 540 B.C., the Ionian cities came under Persian rule, and through this channel the Greeks must have absorbed much Mesopotamian knowledge, even though they did not master cuneiform script. The Greek cities remained as separate states, in contact, in cooperation or in conflict, federating or feuding. Ruled at first by military leaders, civil society was steadily democratized.

In 499 B.C., the Ionians rebelled against their Persian overlords, and appealed for aid to their kinsmen in the mainland state of Athens. Intermittent warfare over a period of fifty years, despite many setbacks, at length secured the Greek cities from further Persian interference. Athens became the major military and cultural center of the Greek world, with a population of perhaps a quarter million (up to half of whom were probably slaves).

It is in fifth-century Athens that we find firm evidence of written records and collections of them. There are abundant indications of the literacy of administrators and scholars, and some for at least minimal literacy among other citizens. For example, inscriptions were posted in public places, intended to be read: wooden boards or tablets were used to bring official notices to the attention of citizens. Further positive evidence comes from the widespread distribution of Greek graffiti—e.g., there are initials of even sixth-century Greek soldiers cut into the statues of Abu Simbel on the Nile. By the fifth century, something like the commercial production of books had developed. Papyrus was almost certainly then available to the Greeks, although the earliest literary Greek papyrus preserved is of the fourth century.

At that time, the following dialogue occurs in a Greek play: "Pick yourself out a nice book. Look at the titles and see if they interest you. Here are Orpheus, Hesiod, Homer. . . . What have you chosen? A cookery book!" Do we have here a bookshop, or a private library? Whichever was the case, there were available multiple copies of the poets, playwrights, historians, philosophers—and gastronomers—of Greece.

EDUCATION, THE ACADEMY AND THE LYCEUM

Earlier, brief mention was made of the education of scribes. Education is the major mode of transmitting knowledge from one generation to the next, and indications of its development will be given from time to time in this book. Greek education in the Homeric age seems to have been little more than the inculcation of military virtues into boys—a type of training kept up longest in Sparta.

The beginnings of intellectual training and the foundations of schools are to be found in the Ionian colonies in the sixth century B.C., while Athenian laws for the regulation of schools, ascribed to the sixth-century Solon, imply that they also existed there in the same century. The pupil was expected to pick up the rudiments of arithmetic, which amounted to reckoning on the fingers or with stones. Gymnastics played a large part in a boy's training (hence the word "gymnasium"), but reading and writing were also taught. The position of girls was generally much less favorable.

For anything in the way of higher education, a man had to seek out someone who would instruct him. The tradition that Thales, Pythagoras and other early philosophers obtained their wisdom in Egypt implies that in their days it was still difficult to obtain any intellectual training in Greece, and that travel to obtain information from other centers of civilization was the only solution. One outstanding traveler who has left us a record of his observations was the historian Herodotus (fifth century B.C.). From his native city of Halicarnassus, in Asia Minor, he visited the Aegean islands, cities in mainland Greece, Macedonia, the coasts of the Black Sea, Persia, Tyre, Egypt and North Africa. He may be seen as an example of a new type of person not found in the older empires—the free (and financially independent) individual inquirer. Once such native philosophers had become established, seekers after knowledge gathered round them—a notable example being the "brotherhood" set up by Pythagoras in a Greek colony in southern Italy in the sixth century.

Toward the end of the fifth century B.C., a considerable demand for higher education arose, to be answered by the appearance of professional teachers known as sophists. Setting themselves up in the marketplace, particularly in Athens, they offered instruction to all who would pay their fees. The popularity of this teaching induced seekers after wealth rather than wisdom to offer their services as sophists, and in the course of time the standard of teaching was lowered.

The sophists—including Socrates—were itinerant teachers. Plato of Athens departed from this practice, setting up a school in a definite location, away from the marketplace. The site he chose was land that belonged to one Akademos, and the name became attached to his school. Plato's interests were more sophisticated than those of the traditional sophist—he taught philosophy, ethics, politics. The Academy continued after his death, undergoing many transformations over the years, and lasted until A.D. 529, when it was closed down by the Christian emperor Justinian. During—and after—its existence, the Academy served as a model for many other schools.

Aristotle had been a member of the Academy before becoming tutor to Alexander the Great, and on the death of Plato he set up a new establishment, called the Lyceum. Here pupils studied philosophy, logic, and science, and conducted research. Alexander on his triumphal marches to the ends of the known world collected and dispatched to his old tutor all kinds of specimens—animals, plants, minerals and other curiosities—thus providing much material for Aristotle and his pupils to work over. Alexander's campaigns might be regarded as incidentally the first scientific expeditions in history.

COLLECTIONS OF DOCUMENTS

The earliest definitely known document collections in Greece were of a kind mentioned by Aristotle in his *Politics*. He refers to an "office that deals with the registration of private contracts and court decisions; indictments have also to be deposited with it." In Athens there have been excavated buildings in the Agora (the central town square) that are known to have been used to house archival records from the fifth century B.C. on. The ruling council of Athens met in an assembly hall, and the building complex eventually housed an archival repository containing laws and decrees, council minutes, financial records and so on. Preservation and consultation facilities must have been efficient, for we know that in 150 B.C. a decree proposed in 300 B.C. was produced without difficulty, and in the third century A.D., material relating to the trial of Socrates was consulted.

In the Aegean island of Cos, during the fifth century B.C., Hippocrates founded a medical school, and writings ascribed to him cover subject matters we now call anatomy, physiology, pathology, hygiene, therapeutics and medical ethics. It is likely that the corpus represents a collection of medical works assembled over a period of time at Cos—the working library of the school. There were no doubt such medical collections in other Greek cities.

The strongest evidence for the existence of collections of manuscripts is the survival of many early Greek writings—though equally as many must have failed to survive. The copious quotations from and comments on earlier writers that are to be found, for example, in Aristotle's works persuasively imply that he had ready access to those writings, even though few texts from his period have been preserved.

The fate of one collection—the writings of Aristotle himself—deserves mention. Writing some hundreds of years after the events to which they refer, the authors Strabon, Athenaeus and Diogenes Laertius left accounts from which may be pieced together the following story.

After Aristotle's death, his papers became the property of his friend and successor at the Lyceum, Theophrastus, who in turn bequeathed them not to his own successor Straton but to his nephew Neleus of Scepsis. The latter does not seem to have cared for them in any way, and his heirs sold some items to Ptolemy Philadelphus, who was building up the library of Alexandria. The rest of the collection was hidden in a cave, apparently to safeguard it against possible seizure by the ruler Attalus, who was building up the rival library of Pergamum. Some time later, a collector called Apellicon obtained the material and added it to his private library in Athens. The Roman general Sulla sacked Athens in 84 B.C. and bought or seized the Aristotelian manuscripts and took them to Rome. In about 70 B.C., the material was worked over and edited by Andronicus of Rhodes, then head of the Lyceum, and it is probably from his edition that all subsequent editions of Aristotle are derived. Copies of his texts spread to other parts of the Greek-speaking world, for example to Constantinople.

Later in this book we will be looking in detail at the ways in which knowledge of Greek writings has come down to us, but as a foretaste it is of interest to continue here the story for Aristotle. The corpus of Aristotelian texts is encyclopedic in scope, covering aspects of logic, mechanics, physics, astronomy, meteorology, botany, zoology, psychology, ethics, economics, politics, metaphysics, rhetoric and poetics. For over two thousand years it has attracted commentators and translators. Early commentators included Alexander of Aphrodisias, head of the Lyceum 198–211, and in the sixth century A.D., John Philoponus of Alexandria. The first Latin commentator was Boethius of Rome (also sixth century).

In the same century, scholars from the Athens Academy moved to Persia, taking with them Greek Aristotelian texts. In the course of time, many of these works were translated into Arabic, and commentaries were written by a series of Arab scholars. In the twelfth century, translation of some texts from Arabic back into Latin was carried out by Gerard of Cremona, but much more extensive Latin translations were provided in the thirteenth century by William of Moerbeke, using Greek manuscripts probably obtained from Constantinople.

In the fifteenth century, many more Greek manuscripts became available from Constantinople, copies dating mostly from the tenth and eleventh centuries. From these, fresh Latin translations were made, largely supplanting the earlier medieval ones and being used for the first printed texts (first edition in Latin 1472, in Greek 1495–98). The standard modern edition in Greek, prepared by Immanuel Bekker, was published in Berlin, 1831–70, with English translation by W.D. Ross in 1908–31. Aristotle lived 384–322 B.C., but he flourished from the fourth century B.C. on!

ALEXANDRIA AND PERGAMUM

We have already mentioned Alexander the Great. His father, Philip of Macedon, established leadership over the Greek city-states, and in 334 B.C. Alexander entered Asia Minor to challenge Persia, an old enemy of the Greeks. In less than ten years the whole Persian empire, from the borders of India to Egypt, fell into his hands. After Alexander's death in Babylon in 323 B.C., his generals vied for the spoils of the empire, the main dynasties founded being the Seleucids, who ruled over the major Asian parts of the empire, and the Ptolemies, who ruled Egypt.

Thousands of Greeks flocked into Asia and Egypt to begin a new era of colonization. The governing administrations were staffed by Greeks, new cities were founded, merchants and artisans followed. Greek culture diffused to the east and south, but equally fresh Mesopotamian and Egyptian influences were brought to bear upon it.

Different areas responded in varying degrees to Greek influence. Hellenized Phoenician cities contributed to philosophy—for example, through the Cypriot Zeno, the founder of Stoicism. In Syria, Antioch became the center of a region referred to as "a new Macedonia." In Asia Minor, flourishing Greek cities developed, such as Pergamum, Ephesus and Smyrna. However, Hellenism was an urban phenomenon—around the cities the uneducated peasantry continued to use the local vernacular, such as Aramaic or Egyptian.

Alexander himself had founded the Greek city of Alexandria in Egypt in 331 B.C. It soon outdistanced all other Hellenic cities as a commercial center. The population eventually reached a million, including Greeks, Jews and Egyptians. The city was no democracy modeled on Athens, but a strongly centralized state owned and ruled by the king, with an official bureaucracy of the ancient Egyptian kind.

Under the royal patronage of the Ptolemies there were created the museum and the library. The museum was part of the royal palace, and its nature was that of a scholarly research institute. Its members lived together like the fellows of a medieval college. It was equipped for astronomical observation, anatomical dissection and physiological experiment, and around it were botanical and zoological gardens. There is no evidence that it was used for teaching purposes, except in the sense of instruction given informally by a scholar to his apprentice assistants. Two pupils of Theophrastus, then head of the Lyceum in Athens, were active in guiding the development of the museum—Demetrius and Straton. Demetrius had been governor of Athens, but moved to Alexandria when Athens was under attack from Macedon. Straton, a younger man, was in Alexandria during 300–288, after which he returned to Athens to take over the Lyceum on the death of Theophrastus.

Linked to the museum by a covered marble colonnade was the Alexandrina or main library. In it were ten halls with armoria (cupboards) for books (in the form of papyrus rolls), each hall assigned to a separate branch of knowledge. An offshoot was the Serapeiana, a daughter library, housed in a temple devoted to the Egyptian god Serapis. Demetrius became the first chief librarian.

After building the museum complex, Ptolemy began to collect scholars, while Demetrius collected books. Altogether probably a hundred scholars were assembled under royal patronage, giving up perhaps some of their intellectual freedom (particularly that of criticizing the institution of monarchy) in return for security from want while pursuing a life of study, and for the enjoyment of a corporate scholarship that must have had much in common with a modern university. In addition to those resident in the museum, there were scholars throughout the Hellenic world who had studied there at some time, like Archimedes, or had otherwise come under its influence. Among the scientists who worked at Alexandria can be named the geometers Euclid and Apollonius; the geographer Eratosthenes; the medicos Herophilus, Erasistratus and Galen; and the astronomer Claudius Ptolemaius (also known as Ptolemy).

Demetrius and his successors, with lavish funds and powers that a modern librarian hardly dare imagine, sent book scouts out over all the known world to purchase where they could and to copy where they could not purchase. It is even recorded that ships entering the harbor of Alexandria were systematically searched for books, which if wanted were confiscated and copied. There are conflicting estimates of the size of the book stock, which may have amounted to half a million rolls.

The research carried out in the museum contributed greatly to the flowering of science and mathematics in the third century B.C. For the first time, there appeared in the Greek world people—such as Archimedes and Claudius Ptolemaius—who were uniquely preoccupied with scientific work. What is also important from the point of view of this book is the great attention paid at Alexandria to the written word. The teaching of the sophists and the Greek philosophers had been by lecture and discussion. The museum began the tradition of book learning and elaborated for the first time the methods of literary scholarship. Variant copies of Greek works were collected and compared with each other, till an edition was prepared which represented the best approach to the original manuscripts. Even when Rome became the main power in the Mediterranean, the museum remained a center of learning. In spite of destruction by fire when Julius Caesar invaded Alexandria (47 B.C.) and by Christian riots at a later period, it was not until the seventh century A.D. that the museum ceased to exist, when Alexandria was taken over by the Arabs.

Alexandria was not without its emulators and rivals. The ruler of the Greek city of Pergamum in Asia Minor, Attalus, around 200 B.C., decided to make his city a seat of art and learning. He and his successor Eumenes (197–159) built up a school and library, the Pergameniana, along lines similar to those of Alexandria. Scholars who were attracted to Pergamum included the mathematician Apollonius. In 41 B.C. the library was plundered by Mark Anthony and the books were presented to Cleopatra for incorporation in the Alexandrian library. As well as at Pergamum, there were created schools and libraries in cities such as Samos, Rhodes, Antioch and Ephesus.

Alexandria played a great part in the organization of the book trade. Books were, as we have indicated earlier, written and copied for centuries before Alexandria was founded, but the large-scale production of texts at the museum must have had a considerable influence on the formation of a regular trade. The city was the home of the best papyrus factories, from where this by-then-universal writing material was exported to Greece and Italy.

An alternative writing material, parchment prepared from animal skin, was developed at Pergamum (according to Pliny) to circumvent the Alexandrian papyrus monopoly. Leather had long been used as a writing material, and it is likely that what took place was an improvement in quality rather than a fresh invention. Certainly the name "parchment" is derived from the Greek *pergamene* (Latin *charta pergamena*), after the supposed town of origin. It was not until well into the Christian era that it displaced papyrus.

THE ROMAN WORLD

Like Athens, the city of Rome was founded during the eighth century B.C., another result of the southward spread of Indo-European peoples. But unlike Athens, Rome did not remain an individual city-state in a loose and unstable federation, but in time imposed its rule over the whole Mediterranean world, extending into Gaul and Britain, and annexing Egypt in 30 B.C.

The unified Roman empire lasted a further four centuries, though subject to increasing pressure from Germanic tribes. At its greatest extent, the empire may have embraced up to 100 million people. Rome itself grew into a city of about a million inhabitants, perhaps a quarter of them slaves.

For the first 250 years of the Christian era, the empire enjoyed internal peace never hitherto existing over such a wide area. Though the empire was predominantly agricultural, cities increased in number, forming nerve centers linked together by a network of roads and waterways. As road builders, the Romans surpassed all earlier peoples, the roads serving for armies, travelers and messengers. It is estimated that by A.D. 300 the grid of paved roads in the empire covered more than fifty thousand miles. They were designed primarily for pedestrians and pack animals, but could carry light springless carriages, and couriers—using relays of carriages or mounts—could, it is claimed, cover 150 miles a day. Craftsmen in pottery and glass migrated across Europe. Goods circulated freely—furs from the barbarian north, spices from Arabia, ebony and ivory from India, even silks from China.

Among the cities overrun by Roman armies were of course those of the Greeks—at first, Greek colonies in southern Italy, then those in mainland Greece, Asia Minor and Egypt. The developed culture of the Greeks captured the Roman mind. Many educated Romans learned Greek, read Greek literature, employed Greek architects and artists. In the third century B.C., a Greek slave Livius Andronicus translated the *Odyssey* and several Greek plays into Latin. Later, Virgil in his *Aeneid* linked Rome to Homer by claiming that the city was founded by the Trojan character Aeneas.

From about 200 B.C., Roman educational practice was built increasingly on Greek models. Some literacy in Latin became widespread throughout the Roman world. The empire needed literate administrators—clerks, registrars and accountants, all performing duties that required at least an elementary education. Schools were introduced into the provinces such as Spain and Gaul—e.g., at Massilia (Marseilles), Arelas (Arles), Burdigala (Bordeaux), Lugdunum (Lyons), and Tolosa (Toulouse). There was a progressive spread of Latin into many parts of the empire. Often it shared the field with the earlier language of a region (e.g., Greek), but in Gaul and Spain it eventually replaced the earlier Celtic languages.

One evidence of appreciable literacy in Rome itself is *Acta diurna,* an official public gazette instituted by Julius Caesar in 59 B.C. It contained the proceedings of the Senate, law reports, imperial enactments and items of general news. Copies were available in the bookshops that developed at that time.

BOOKS AND LIBRARIES

Books were produced on papyrus on all manner of topics, not only literary genres such as poems and plays. Historical and philosophical writings were common, as well as biography and collections of letters such as those of Cicero. Technical treatises—particularly on agriculture, architecture and medicine—were also published, as were handbooks of instruction.

Mumby has pictured the publishing activities of Rome in the following words:

> We stand in the Argiletum thoroughfare. The pillars outside the shops are covered with the titles of the works to be obtained within, and the whole place is evidently a favourite haunt of literary men. . . . If we look within the larger publishers' offices we shall find slaves hard at work on a new edition of the latest book. . . . We are surprised by the marvellous cheapness of the books—often no more than a few pence. . . . Slave labour, with all its drawbacks, made it possible for Rome to manage remarkably well without the printing press.

The first publisher whose name has come down to us is Atticus, a friend of Cicero (first century B.C.). This wealthy man dealt in books on a grand scale, employing large numbers of slaves and freedmen. Not only a publisher, he was also a retail bookseller, selling single books or whole libraries. One of his products, Varro's book on portraiture, required equipment for the mechanical reproduction of its hundreds of pictures.

From such sources wealthy individuals built up private libraries in their villas. Atticus himself had a library of twenty thousand rolls, and Cicero wrote to him: "Mind you don't promise your library to anybody, because I am hoarding up all my savings to get it as a resource for my old age." In the first century A.D., Seneca wrote that "libraries are as necessary in the homes of the rich as baths with hot and cold water." Private villa libraries were probably the main channel through which the texts of classical Latin were handed down, till they eventually found a precarious refuge in monastic libraries.

Irwin has brought together a good deal of evidence for publicly available—or at any rate, publicly maintained—libraries in Rome and the empire. In Rome itself, probably the most important was the Ulpian, established by the emperor Trajan in A.D. 100, to which a reference is still found as late as the fifth century. There is evidence of publicly maintained libraries in provincial cities such as Tivoli, Como, Athens, Ephesus, Tripoli, Carthage and Timgad (in North Africa), and it is likely that libraries also existed at places that were important centers of Roman education, such as Ravenna, Toulouse, Bordeaux, Marseilles and Beirut.

It is convenient to mention here a change in the materials used for writing that occurred during the later Roman period. After the third century A.D., there is steadily less evidence of papyrus, though the last recorded use of it was not until the eleventh century. Parchment began to replace it. Moreover, from the first century A.D., or perhaps even earlier, the roll form of book (whether papyrus or parchment) began to be challenged by a new form, a collection of leaves fastened together to form a codex, essentially the same construction as the modern book. From the fourth century, this became the dominant form.

ENCYCLOPEDISTS

Though there is little Roman writing that contributed with originality to science, the communication of knowledge was aided by its becoming increasingly codified and systematically organized. Handbooks summarizing existing knowledge expanded into lengthy treatises, of which the first of significance were those of Varro in the first century B.C. He wrote many books, few of which have survived. One lost treatise was the *Nine Books of Disciplines*, covering grammar, dialectics and logic, rhetoric, geometry, arithmetic, astrology and astronomy, music, medicine and architecture, which was much used by later compilers. In the same century, Vitruvius wrote a book on architecture that in fact ranged widely over the science and technology of his time, citing many Greek and Latin sources.

In the first century A.D., the major encyclopedic work was that of Pliny the Elder. Of the original 102 books of his *Natural History*, 37 have survived, the summary of a lifetime of reading, written with the aid of an army of Greek clerks. It was a mine of historical information compiled from about two thousand sources, most of which have been lost. Pliny had little critical ability, transmitting truth and error with equal zeal. Gibbon aptly described his work as an "immense register in which are deposited the discoveries, the arts, and the errors of mankind." It became the most widely read secular book of Christian Europe.

Contemporary with Pliny was the physician Dioscorides, who wrote a Greek treatise, translated into Latin as *Materia medica*. This dealt mainly with plants and their medicinal uses. The plants were grouped into families, each species was carefully described, and directions for preparing herbal remedies were given. This book was the direct ancestor of the medieval herbals, and later served as a starting point for Renaissance botanists. During the middle ages, illustrations were added to Dioscorides' text—drawings probably derived from those of the physician Crateuas in the first century B.C. Through constant copying, the illustrations steadily worsened. Even in Pliny's time, he was led to comment, "Pictures are very deceitful. . . . No marvel if they that drew them did fail and degenerate from the first pattern and original."

Parallel with this botanical text was the second century A.D. work *Physiologus*, the model for all later medieval bestiaries, again enormously popular. We know of versions in Greek, Latin, Syriac, Arabic, Ethiopian, Armenian, German, Anglo-Saxon, Icelandic, Rumanian and other languages.

THE END OF THE WESTERN EMPIRE

As early as A.D. 150, Roman society began to decline. As craftsmen migrated to outlying provinces, there was a reduction of interprovincial trade. Each province tended to become once more a self-sufficient economic unit. Farms began to grow into self-sufficing households, forerunners of the medieval manors. The market for the products of city-based manufacturers contracted. The once-flourishing cities were impoverished and actually shrank in size. By A.D. 250, prosperity had vanished. Economically, Roman civilization began to collapse long before barbarian invaders from Germany disrupted the political unity of the empire.

The invasion of the empire by Teutonic tribes was a constantly recurring event from the third century onwards. Many tribes were absorbed peacefully. German soldiers served in Roman legions, some of them reaching high rank—the commander in chief of the army of the fourth-century emperor Honorius was a Vandal, Stilicho.

The German invasions were in part the consequence of a great movement of peoples that originated earlier on the other side of the world. It is often said that the Great Wall of China, built about 200 B.C., deprived the neighboring Huns access to the rich grass plains of that country. Generations of war between the two peoples ended in a disastrous defeat of the Huns by the Chinese in A.D. 93. Whether for this reason or because climatic changes reduced their food supplies, the Huns gradually migrated westward to the Volga river, crossed it in 374, and eventually reached the Danube.

The peoples whose lands the Huns invaded were set on the move. Among them, the Visigoths, who had lived first on the Baltic, then north-west of the Black Sea, fought their way through Greece and the Balkans, and into Italy. Led by Alaric, they captured and pillaged Rome in 410. The Vandals also came from the north of Europe, and moved through Germany and France to Spain. In 427 they crossed into Africa, advanced along the North African coast, and took Carthage in 439. In 455, they invaded Sicily and sacked Rome, which a few years earlier had been in danger from Attila the Hun himself. Finally, in 476, Odoacer, the most powerful of the German generals in Italy, banished the reigning emperor, and the Roman empire came to an end in the west. The invasions of the Germanic peoples, who were unused to cities and ignorant of advanced technology, submerged the declining civilization of Rome.

Section 3

The Medieval Period (500–1450)

TRANSMISSION OF CLASSICAL KNOWLEDGE

It is historically inept to characterize the medieval period—say, from the end of the Western empire in 476 to the invention of printing in about 1450—as concerned solely with the transmission of a cultural heritage. The medieval world, both Christian and Islamic, itself made many contributions to culture, including some of man's most beautiful buildings. But from the vantage point of contemporary science, the work of both Christianity and Islam in preserving and transmitting the cultural heritage looms large, and will be emphasized in this book.

Not that the transmission was in any way complete. As Diringer puts it:

> Many peoples and events have conspired to deprive us of the greater part of the literary treasures of antiquity:
>
> - whether it be the Persians, who delivered a fatal blow to Phoenician and Egyptian literatures, when in the sixth to fourth centuries B.C. they destroyed the Phoenician temples and schools, and the Egyptian temple-colleges;
> - whether it be political persecution, not unknown even in ancient times, as exemplified by the public burning in 411 B.C., at Athens, of the books of the sophist and grammarian Protagoras; or by the public burnings of books in Rome, even by enlightened emperors such as Augustus, who allegedly confiscated and burned 2000 books;
> - whether it be the barbarians of the north, who destroyed the famous Roman libraries;

Figure 6 Medieval Centers (Europe and Islam)

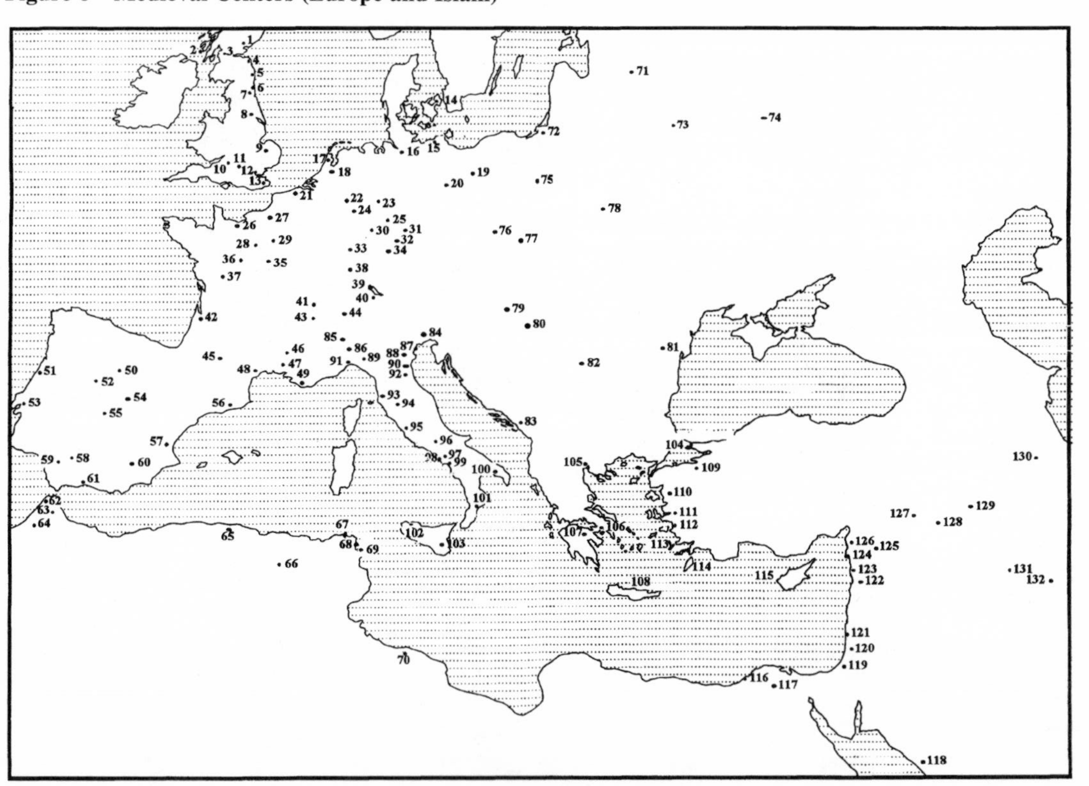

Aachen: 22
Aleppo: 125
Alexandria: 116
Algiers: 65
Amsterdam: 17
Antioch: 126
Arles: 47
Athens: 106
Augsburg: 34
Avignon: 46

Baghdad: 131
Bamberg: 31
Barcelona: 56
Basel: 38
Bath: 10
Beirut: 123
Belgrade: 82
Bobbio: 89
Bologna: 92
Bordeaux: 42
Bruges: 21
Bucharest: 81
Budapest: 80
Bursa: 109
Byzantium: 104

Caen: 26
Caesarea: 121
Cairo: 117
Cambridge: 9
Canterbury: 13
Carthage: 68
Catanzaro: 101
Chartres: 28
Cluny: 41
Coimbra: 61
Cologne: 24
Constantinople: 104
Copenhagen: 14
Cordova: 58
Corinth: 107
Corvey: 23
Cos: 113
Cracow: 77
Crete: 108
Cyprus: 115

Damascus: 122
Durham: 7

Edessa: 127
Ephesus: 112
Erfurt: 20

Fez: 63
Florence: 94
Fulda: 25

Gaza: 119
Geneva: 44
Genoa: 91
Glasgow: 3
Granada: 60
Gundisapur: 132

Hamburg: 16
Hippo: 67

Iona: 2

Jarrow: 5
Jerusalem: 120

Kiev: 78
Konigsberg: 72

Latakia: 124
Leipzig: 19
Lindisfarne: 4
Lisbon: 53
London: 12
Lyons: 43

Madrid: 54
Mainz: 30
Malaga: 61
Marrakesh: 64
Marseilles: 49
Medina: 118
Milan: 85
Monte Cassino: 96
Montpellier: 48
Moscow: 74
Mosul: 129

Naples: 98
Nisibis: 128
Nola: 97
Novgorod: 71
Nuremberg: 32

Orleans: 35
Oxford: 11

Padua: 88
Palermo: 102
Paris: 29
Pavia: 86
Pella: 105
Pergamum: 110
Pisa: 93
Poitiers: 37
Prague: 76

Ragusa: 83
Ravenna: 90
Reichenau: 39
Rhodes: 114
Rome: 95
Rostock: 15
Rouen: 27

St. Andrews: 1
St. Gallen: 40
Salamanca: 52
Salerno: 99
Seville: 59
Smolensk: 73
Smyrna: 111
Strasbourg: 33
Syracuse: 103

Tabriz: 130
Tangier: 62
Tarentum: 100
Timgad: 66
Toledo: 55
Toulouse: 45
Tours: 36
Treviso: 84
Tripoli: 70
Tunis: 69

Utrecht: 18

Valencia: 57
Valladolid: 50
Venice: 87
Vienna: 79

Warsaw: 75
Wearmouth: 6

York: 8

- whether it be the zeal of the Christians, who at different periods made great havoc of non-Christian or "heretic" works;
- whether it be the Mongols who ravaged the books of the Moslems, or the Arabs who ravaged the books of the Christians;
- whether it be the ravages of time and climate, of damp or the bookworm, or of fire. . . .

The wonder is, not that we have so few ancient writings in our present possession, but that we have any.

The story of medieval transmission is complicated because classical Mediterranean culture diverged into three streams that came together again only at the end of the period. An Eastern empire, centered on Constantinople (Byzantium), survived the sack of Rome and preserved Greek culture, though at a low level of vitality. The Arabs overran the southern shores of the Mediterranean and absorbed part of Greek culture. On the mainland of Europe, Christianity kept Latin learning alive, and eventually came into contact with both Arab and Byzantine culture. The three streams will be considered in turn.

THE EASTERN EMPIRE

The Roman empire was partitioned in A.D. 396, the eastern half then including Greece, the Balkans, Asia Minor, Syria and Egypt. For a time—under the rule of Justinian in the sixth century—Italy, North Africa and even southern Spain were under Eastern control, but the territory shrank under the attack of Germanic tribes, Slavs and Arabs. By the twelfth century it was closely concentrated around Constantinople, and this city was conquered by the Crusaders in 1204.

During the early part of the Eastern empire's existence, there is evidence of active cultural life at a number of centers besides Constantinople, such as Antioch, Beirut, Gaza and Edessa. Attestation for large book collections in Constantinople is provided by the citations of compilers such as Photius (ninth century) and Tzetzes (twelfth century), and by the survival of ninth-century manuscripts from the library of one Arethas. However, the burning and pillaging of the city by the Crusaders caused great loss of texts.

The culture of the Eastern empire was essentially Greek, but with the adoption of Christianity by the emperor Constantine in the fourth century, learning increasingly came under the control of the church. Aristotelian logic and Platonic idealism came to be used to expound, amplify and defend Christian doctrine. The inherent difficulties of this task inevitably provoked controversies that split the church into sects. Even before the growth of external attacks on the empire, these schisms had a weakening and disintegrating effect.

One schism relevant to our story occurred in 431. In that year, a general council of the church condemned and excommunicated a certain Nestorius, Patriarch of Constantinople. Nestorius came from Antioch, in Syria, and many Syrian Christians repudiated the general council. During the next half century, the "heretics" were persecuted by the orthodox, and many migrated eastwards towards Persia and beyond—some even reached China.

The Persians were never subjugated by the Romans, and in 226 had founded a new Sassanid empire. Collisions between the two powers were frequent, and after the division of the Roman empire, friction between Persia and Constantinople continued. The fact that the Nestorians were a dissident minority made them welcome in Persia. The sect was encouraged to set up a school at Nisibis in Mesopotamia, which was later transferred to Gondisapur (Jundishapur) near the Persian Gulf.

The Nestorians, as has been said, were largely Syrian. Their language, Syriac, was a branch of Aramaic, which had been the lingua franca of the whole region since the sixth century B.C. It was the principal speech of traders from Egypt and Asia Minor to India. In the fifth century A.D., Syriac was (after Greek) the most important language in the Eastern Roman empire, and it was therefore natural that many Greek works should be translated into Syriac. One of the most notable Nestorian scholars was Sergius of Resaina, who translated many works by Aristotle and Galen, as well as other medical and agricultural treatises. There is extant from the period in which he lived—the first quarter of the sixth century—a *Book of Medicines*, a compilation of Galenic writings.

In 529, the emperor Justinian, as a Christian, forbade the further teaching of pagan philosophy in the empire, including Athens, and closed the Academy that had been founded by Plato. Simplicius and six other scholars from the Academy for a time took refuge in Persia. While they were there, Chosroes came to the throne of that country. Chosroes greatly admired Greek culture, and despite (or because of) the war then going on between Persia and the Eastern empire, he established an academy at Gondisapur, linked to a hospital, medical school and observatory. Nestorian scholars directed the academy, and the teaching was in Syriac.

Gondisapur in this way became a center in which Greek science was preserved, translated and taught, mingling with the native learning of Persia itself. The academy received other influences as well: Chosroes sent one of his doctors, Burzuya, to India, from which he brought back many medical books. When Persia was conquered by the Arabs in 652, the work of the academy did not end. On the contrary, it developed and flourished. In due course, the Abbasid caliphs (rulers of the eastern part of the Arab empire) took over the patronage of the scholars of Gondisapur, and the Greek heritage was transmitted to the Arabs.

TRANSMISSION INTO ARABIC

In the seventh century, a new force appeared in the world: the Arab peoples, inspired by the faith of Islam, began their career of conquest. It is frequently alleged that Islam arose in a primitive community of desert nomads, but this is far from the truth. As O'Leary has pointed out, Islam had its origin in a district into which advanced civilization had percolated from remote ages. Arabia had never been fully part of the civilized world, but it hung on the fringe, not isolated but on the edge of the cultural life of the Mesopotamian and Mediterranean peoples.

The country lay across the route to India, and Arabs in fact dominated trade between India and Europe even in Roman times. The geographer Strabon in the first century B.C. compared the size of one of their caravans to that of an army. This commercial intercourse, though it did not at first lead to any marked cultural development, must undoubtedly have helped to make the Arabs receptive to foreign ideas. It is at any rate a fact that their interaction with the remnants of the Greek and Persian civilizations was very fruitful.

In the century that followed the Hegira (622, when the prophet Muhammad established his position at Medina), the Arabs conquered one by one the lands bordering the Mediterranean (Syria, Palestine, Egypt, the whole North African coast, and eventually Spain), thus overrunning towns that had begun as Phoenician, Greek or Roman colonies. In 669 they laid siege to Constantinople and did not finally give up their attempts to subdue the city for fifty years. Penetrating into France, they were at last repulsed in 732 at Tours and Poitiers. They traveled eastwards through India to China, where they successfully fought a battle at Talas in 751, but did not advance further.

The Arabs thus became masters of a huge area in which centers of Greek, Persian and Hindu learning had existed for a thousand years, and once the period of turbulent expansion was over they quickly made this rich heritage their own. As we have noted, the academy at Gondisapur continued under Arab rule, and Alexandria was taken in 640. Here there had been preserved the last vigorous remnant of Greek science—the Byzantine school of medicine. This had flourished at Constantinople since the time of Oribasius in the fourth century, at which time a commission of doctors had prepared a "standard edition" of Greek medical works—twelve books of Hippocrates and sixteen of Galen. The tradition was maintained in the sixth century by Aetius of Amida and Alexander of Tralles. The last important writer was Paulus Aegineta, who remained in Alexandria after the Arab conquest. In 718, the school was transferred to Antioch, and in later years to Baghdad.

The Abbasids in 762 had founded Baghdad as a new eastern capital city. The caliph al-Mansur invited to it scholars from neighboring cities, including Gondisapur and Antioch. His successor, Harun-al-Rashid, took an even greater interest in science and literature. He sent out agents to purchase Greek manuscripts in the Roman empire, and in his reign Baghdad became a center for the translation of Greek works into Arabic. Among the first areas to be explored were geometry and astronomy, and translations were made of Euclid's *Elements* and Ptolemy's astronomy, which became known by its Arabic title of *Almagest*. In the same period, al-Batriq and his son, Yahya ibn al-Batriq, translated works by Galen, Hippocrates and Aristotle.

The most celebrated translator was Hunain ibn Ishaq. He was the son of a Nestorian druggist, a Syrian, and studied medicine at Gondisapur. He traveled, learned Greek and later Arabic, and eventually set up in Baghdad as a translator. The caliph al-Mamun founded in about 826 a "House of Wisdom" as an institution for preparing translations, with Hunain in charge, and work went on steadily. By 856, he and his collaborators had translated ninety-five books of Galen into Syriac and thirty-nine into Arabic, making available the whole Alexandrian "canon" of Galenic medicine. His son, Ishaq ibn Hunain, translated works by Aristotle, Euclid, Archimedes, Ptolemy and others.

With the achievements of the House of Wisdom at the end of the ninth century, the Arabs began to pass from simple translation to the production of original works. Hunain himself wrote *Ten Treatises on the Eye*, an early textbook on ophthalmology. Thabit ibn Qurra and his school not only carried out further translations, but are credited by a later writer with 165 works on logic, mathematics, astronomy and medicine, in Arabic and Syriac.

Thus during the period 750 to 900, the scholars of Gondisapur and Baghdad—many of them Syrian or Persian, not Arab, and Christian, not

Moslem—made available a large part of Greek science to the Arab-speaking world (though not Greek drama and poetry). Mieli estimates that the Arabs came to know the whole of Greek mathematics, though they failed to appreciate fully the work of Archimedes. Greek mechanics (Philo and Heron), astronomy and mathematical geography (Ptolemy) were thoroughly studied and developed. The Arabs were conversant with all the main works of Greek medicine, although they were more influenced by the Galenic than by the Hippocratic tradition. The botanical work of Dioscorides was influential, but they seemed to have had no knowledge of the botanist Theophrastus. They knew Aristotle's zoological writings, but did not follow up his advances. Latin writers such as Varro, Pliny and Vitruvius were completely unknown.

CENTERS OF ARAB SCHOLARSHIP

This, then, was the heritage of learning that existed in Arabic by about the year 900. What use was made of it? The great historical merit of Arab science is to have preserved the knowledge of the Greeks, so that it was readily available when the reviving scholarship of Europe came into contact with the Arabs in the eleventh century. This preservation was not a passive affair. Arab scholars worked over the translations they or others had made, preparing commentaries on them, digests and compendia, developing and amplifying classical theories. Arab presentations of knowledge were often, writes Mieli, clearer, better constructed and more subtle than their Greek originals, though they were generally much more prolix. Moreover, in the great mass of writings that they left, the Arabs recorded not a few original observations and theoretical advances.

A word should be said about the connotation of "Arab science." The greater part of the works composed within the Arab empire were written in Arabic, and it is from this fact that the period here dealt with is so named. But the most prominent scholars were rarely Arab by race—they included Egyptians, Jews, Spaniards, and above all, Persians. Some of them will be named when we later consider how Arab science influenced western Europe.

The Arabs (in this wide sense) made many advances in science and technology. They developed algebra, built observatories. In medicine they took further the work of the Greeks, and in chemistry (not developed in Greece but studied in India) they made a good beginning. In manufactures they outdid the world in variety and beauty of design and workmanship. They worked in all the metals—gold, silver, copper, bronze, iron, steel. In textile fabrics they have never been surpassed. They made glass and pottery of the finest quality. They made tinctures, essences and syrups. They made sugar from cane, and prepared many fine kinds of wine. They had good systems of irrigation, and knew the value of fertilizers. They introduced into the West many trees and fruits from the East, and wrote scientific treatises on farming.

A knowledge of papermaking came to the Arabs from China during the eighth century, after a battle in central Asia. Chinese prisoners were taken, and they set up a workshop in Samarkand, the process later being transferred to Baghdad. It is claimed that by 900 there were over a hundred shops in that city employing scribes and binders to produce books for sale.

The development of Arab scholarship began, as we have seen, with the founding of Baghdad in 762. The city became famous for its numerous colleges and libraries, for its translating and teaching. The House of Wisdom was in many respects the most important scholarly institution since the Alexandrian Museum. The Abbasid dynasty ruled Baghdad till it was sacked by the Mongols in 1258. At that time, the scholar Nasir al-din al-Tusi, who accompanied the Mongol armies, saved many books from the plunder, and from books collected in Syria, Iraq and Persia he later built up at Maraga in Azerbaijan his own library (unconvincingly alleged to be four hundred thousand items).

Arab learning was not confined to Baghdad. The Samanid dynasty, that ruled in Bokhara from 872, established a splendid library, which described by the Persian medico Ibn Sina (Avicenna) two centuries later: "I entered a mansion with many chambers, each having chests of books piled one upon another. . . . Each apartment was set aside for books on a single science. I glanced through the catalogue of the works of the ancient Greeks, and asked for those I required." To suggest the alleged size of private libraries, we may note Gibbon's anecdote of a doctor who refused an invitation to reside with the sultan of Bokhara, because the carriage of his books would require four hundred camels.

In 827, Sicily was invaded by the Arabs, and finally conquered fifty years later. This country, too, became a center for the production of works in Arabic, and many of its manuscripts are now in the libraries of Palermo and Naples.

The Fatimid dynasty was established in Africa in the tenth century, and founded Cairo in 968. The royal library in that city, claimed Gibbon, "consisted of 100,000 manuscripts, elegantly transcribed and splendidly bound, which were lent, without jealousy or avarice, to the students of Cairo."

In Europe, the greatest center of Arab science and art was probably Cordova. Here, as in other Spanish towns such as Seville, Malaga and Granada, institutions were founded that have been likened to universities, and libraries flourished. The rulers of Cordova initiated searches in Constantinople for new Greek manuscripts. The Cordovan library is (unconvincingly) alleged to have contained four hundred thousand (or even six hundred thousand) volumes, "44 of which were employed in the mere catalogue." Preeminent among the scholars who worked at Cordova was Ibn Rushd (in the twelfth century), known in Europe as Averroes. His commentaries on Aristotle, later translated into Hebrew and Latin, were of great philosophical influence in later centuries. Like Avicenna of Bokhara, on the other side of the Arab empire, who wrote the *Canon* of medicine, Averroes wrote a medical encyclopedia, translated into Latin as the *Colliget*, that became equally influential.

These two scholars illustrate a basic feature of the Arab world. Centers of learning and libraries were created all over the empire. Where scholars went, their books went with them. At every center, translations from Greek science and Arab works developing from them were to be found, and many would be available for Latin translation in the next stage of the transmission of Greek science.

By the thirteenth century, Arab science was in decline, although it continued to flourish in the East for another two centuries. The center of scientific advance had shifted to western Europe, and it is to this region we must now return, going back to the point in time at which we left it in an earlier section.

THE BARBARIAN WEST

By A.D. 500, Europe was divided into a number of Germanic kingdoms. To the northwest, in what is now France and Germany, were the Franks and Burgundians. To the southeast of them, in Italy, were the Lombards and the Ostrogoths. To the southwest, in modern Spain, were the Visigoths. To the northwest were the Saxons, who with the Angles had begun to invade Celtic Britain. This book will not attempt to trace the way in which these tribes developed into the states of Europe that we know today, although attention will later be given to the growth of vernacular languages.

The "barbarian" conquerors of Rome were peoples only just abandoning nomadic life, their agricultural technique still primitive, and they were reluctant to live in towns, many of which became deserted. There disappeared from Europe everything dependent on large-scale operation: communications, long-range trade, waterworks and so on. Although no less intelligent and vigorous than the peoples they conquered, the barbarians were forced by circumstances to confine their attentions to the more elementary material needs of life.

Nevertheless, they were not so primitive as was once commonly believed. They brought to western Europe such widely divergent gifts as furs and trousers, the art of building houses that were more suited to the climate of northern Europe than the Roman patio type, feltmaking, cloisonné jewelry, the ski, the making of barrels and tubs, new grains like oats and rye, hops, and the art of falconry.

In fact, the so-called Dark Ages of Europe (A.D. 500–1000) began to make progress in applying productive techniques that had previously been restricted in use because of the supply of cheap slave labor. The Christian monasteries, which were founded all over Europe from the sixth century on, did more than tenuously preserve the tradition of book learning. They were also the centers of a systematic drive to clear the land of forest and to farm it more efficiently; they were the home of crafts such as masonry and glass-making.

Europe in the Dark Ages was almost exclusively agricultural, and it is in this field of knowledge that the most decisive technical advances were made. In the ninth or tenth century three inventions were made that greatly increased the pulling power of draft animals: the classical yoke, with its strap pressing on the animal's neck, was replaced by the modern horse collar, which rests on the shoulders and allows the animal to exert its full weight in pulling; the tandem harness enabled a number of animals to be used together; and the metal horseshoe gave adequate protection on rough ground. Such inventions were not made by scholars, but by unlettered craftsmen, and no contemporary writings describe them.

Developments occurred in engineering as well as agriculture. Two new sources of power were widely applied. The waterwheel, which had been known to the Romans but not much used, rapidly became established in Europe from the fourth century onwards. It was used to grind corn and to move the carpenter's saw and the smith's bellows. As Europe emerged from the Dark Ages, we find the earliest example of a windmill, built in Normandy in the twelfth century. These new sources of power led to the mechanization of several industrial processes, such as fulling, which involved beating cloth in water: mechanical hammers replaced clubs wielded by hand. Again, the hotter blast obtained from mechanical bellows permitted the development of cast iron. Perhaps the most spectacular results of medieval engineering are to be seen in the large churches, the design of which required the solution of quite new statical problems.

CHRISTIANITY AND CLASSICAL CULTURE

In western Europe, after the collapse of the Roman empire, the preservation and development of learning was almost exclusively the province of the Christian church, which provided for education, administration, and even law and medicine. Christianity had spread rapidly from its place of origin—north through Asia Minor to Greece and Rome, south to Alexandria. By the end of the second century it was established in Lyons, in Carthage, and perhaps in Spain and Britain. It became organized into urban territories, each a diocese with its bishop and church. The writings of the early churchmen strongly indicate that they had access to, and made good use of, collections of Latin (and sometimes Greek) classical texts. Early in the third century, Clement, working in Alexandria, cited many hundreds of authors. In the following century, both Eusebius and Jerome made mention of the church library at Caesarea. It is also clear that Augustine, in fourth-century Carthage, had access to many classical books.

In Europe, as towns declined, new centers in which learning might be concentrated developed. The earliest Christian monasteries seem to have been in Egypt, perhaps in the second century, but in Europe they are first found in the sixth century. An early example was the monastery at Vivarium in the south of Italy, founded by Cassiodorus in about 540. He seems to have quite consciously seen the need for an institution that could preserve learning in a time of war and devastation. He acquired codexes from all over Italy and from North Africa. He organized more translations of Greek philosophy and science into Latin, and insisted on meticulous care in the copying of texts. According to Thompson, his library certainly contained works by Homer, Hippocrates, Galen, Plato, Dioscorides, Aristotle, Euclid, Archimedes and Ptolemy, as well as Latin authors such as Lucretius, Varro, Pliny and others. A friend of Cassiodorus was Boethius (480–524), who may be called the last great scholar of the classical period. He cherished the plan of translating all Plato and Aristotle into Latin, though little of his planned productions remain. He certainly did provide all that the Dark Ages were to know of Aristotle's logic, and wrote on arithmetic, geometry, mechanics, astronomy and music. Another widely read author of the same period was Isidore, bishop of Seville, whose *Etymologies*, a popular encyclopedia, cited many classical authors.

Of greater ultimate effect on literary transmission was the monastic movement founded in 529 by Benedict at Monte Cassino, near Naples. The rules that he laid down for his flock included instructions as to the regular reading of books to be obtained from the monastery library. This insistence on reading led to the practice of copying texts to provide reading matter, and a "scriptorium" became a regular feature of many Benedictine monasteries. Much of what was read and copied was of course theological rather than scientific, and transmission or even creation of literature was only incidental to monastic life, but these small centers of learning were nevertheless the mechanism by which much classical culture was transmitted. They formed islands of learning in a sea of lay illiteracy.

In the larger monastic (and, later, cathedral) establishments the production of manuscripts was thoroughly organized. The writing was usually done communally in a scriptorium furnished with desks and cupboards (*armaria*) for the writing equipment and for the storage of books. In charge was the *armarius*, who provided the parchment, pen, ink, and other paraphernalia, and assigned work to the transcribers. The more skilled of these (*antiquarii*) copied devotional works and such favorites as Pliny or Boethius, while the less skilled *librarii* carried out preparatory tasks—ruling up parchment, making ink, etc.—and copied secular works. A corrector compared copies with originals and made necessary amendments. Where appropriate, the manuscript was then handed over to the *rubricatores*, who filled in the large initial letters in spaces left for them, and to the specialists in ornamentation, the *illuminatores*. These might be laymen, skilled craftsmen living within the monastery precincts. Bookbinders finished the job.

A certain amount of trading between monasteries took place, later also with cathedral schools. There was interlending, but it was carefully controlled because of the scarcity of books, often requiring a pledge to the value of the loan.

In England, Benedictine monasteries were founded at Wearmouth in 674 and Jarrow in 682. Their bishop (named Biscop) traveled frequently to the Continent, collected manuscripts of the classics as well as ecclesiastical literature from Italian and Gallic monasteries, and introduced craftsmen such as *illuminatores* from France. The greatest figure in European learning at this time was Bede, who divided his time between the two monasteries, though chiefly at Jarrow. Here he gave lessons on the calendar and astronomy. His course was extended to cover such scientific topics as the winds, tides, weather and animals, his chief sources being Pliny and Isidore.

The first result of Bede's teaching was the development in Britain of episcopal or cathedral schools. Schools were founded by his pupils at York and Canterbury. Alcuin, who headed the school at York, has left a detailed account of teaching there. The subjects taught were the seven "liberal arts"—grammar, rhetoric, dialectic, arithmetic, geometry, music and astronomy—plus law, natural history and, of course, the scriptures. Alcuin was soon to leave England and become a great educational reformer at the court of Charlemagne.

In the eighth century, Charlemagne had widely extended the kingdom of the Franks, and promoted church-building, agriculture, manufacture, commerce and education. At his palace in Aachen, a school had been in existence since the sixth century to train youths for the priesthood and for royal administration. Charlemagne (who himself struggled gallantly to learn to read and write Latin late in life) reorganized the teaching, and laymen were allowed to attend the school. In 782, Alcuin was called from York to take over the school and to advise on all matters of education.

One department of learning to feel the reforming hand of Alcuin was the monastic scriptorium. He found that the copying of manuscripts was being neglected. Each scriptorium had developed its own variant script. Alcuin reformed and standardized the script, adopting a particularly clear, neat book-hand that he ordered to be used in all the scriptoria in the kingdom. Subsequently known as the Caroline miniscule, it later became a model taken up by the humanists of the Italian Renaissance, and their book-hand in turn formed the model for the early printing types known as Roman, of which the type used in this book is a descendant. The monastery schools throughout Charlemagne's domain were reformed. During the ninth century, such monasteries as Fulda, Corbie, Reichenau and St. Gall became important centers of learning with famous scriptoria and collections of manuscripts.

The next stage of development was centered on the cathedral schools and their libraries. The great stone cathedrals of Europe were mostly built between 1050 and 1350, soaring to heights not seen since ancient times, symbols of the changing conditions of European life. Cathedrals were far more than merely large churches; they were something like religious universities, where training for the priesthood took place and often secular training as well. The cathedral libraries often had more funds than the older monastery libraries for the acquisition of books, and contained more secular books. Examples are those at York, Durham and Canterbury; at Notre Dame and Rouen; at Bamberg and Hildesheim; and at Toledo and Barcelona.

The greatest French-cathedral book collection of the eleventh century was at Chartres. A school was founded there in 990 by Fulbert, and he set about improving the meager collection of books available at that time. Some record survives of the volumes he added: in logic, the *Categories* of Aristotle and writings by Boethius and Cicero; arithmetic was represented by Boethius and by later writers such as Alcuin; and the astronomical works of Bede were added. After his death, the library continued to grow, largely from donations. For example, in 1144 Herman Contractus sent from Toulouse a copy of Ptolemy's *Planisphere*, which had recently been translated from Arabic into Latin.

To continue the story a little further, in other libraries of the twelfth century were to be found the Latin agricultural texts of Columella and Palladius, Vitruvius on architecture, and the geography of Mela. Durham cathedral library in 1162 contained several medical translations from the Arabic, including Hippocrates and Galen. It is clear that by this time European knowledge of classical science was beginning to enter a new phase—its recovery from the Arabic.

We may end this section by noting here the development of the first "union catalogue," listing the holdings of 183 British church libraries: this *Registrum librorum Angliae* seems to be from the thirteenth century.

CRAFT KNOWLEDGE

In the eleventh and twelfth centuries, crafts and technical accomplishments, exhibited particularly in the building of the great cathedrals, began to come within the province of guilds. These organizations had existed in Roman times primarily as religious and fraternal groups, and when they made closer connections with trade, commerce and crafts they retained their religious character. The masons who were responsible for the construction of cathedrals used their craft lodges, built alongside the cathedrals, as meeting places where they conducted fraternal rites. In time, trade skills became part of these rites, secrets that were jealously guarded. Closed brotherhoods were formed, concerned chiefly with preserving the secret skills of their occupations. So developed a system of qualifications for admission to the guilds, which may be seen as a technical counterpart to that developing simultaneously, but separately, in the cathedral schools.

Among the crafts that formed guilds were cordwainers (shoemakers) and cobblers, pewterers, carpenters, saddlers, bowers, spurriers, ironmongers, founders, coopers, cutlers, goldsmiths, potters, candlemakers and leather tanners. A whole series of crafts dealt with textiles and their manufacture into clothes: weavers, fullers, shearers, dyers, drapers, tailors. Victualling crafts included fishmongers, grocers, vintners, bakers and cooks. The masons, of all medieval craftsmen, left the greatest and most enduring legacy of their work: cathedrals, monasteries, castles, royal strongholds, colleges, hospitals, parish churches, town walls, guild halls, quays and bridges. These structures required quantities of stone, mortar, lead, timber and iron, supplied by other trades.

The craft guilds became the means by which medieval Europe maintained and developed its technical learning. By diligent study, they revived practices of antiquity that had been lost, and learned new techniques that had been developed by the Arabs. Perhaps the most dramatic acquisition was that of gunpowder. Less conspicuous but even more significant were those inventions directly concerned with literacy: papermaking and block printing. By the twelfth century the papermakers were an established guild; the scribal guilds were using block printing for rubrics. In time, the guilds developed a literature of their own, often written in a local vernacular.

One such product that has survived is the sketchbook of the thirteenth-century French engineer, Villard de Honnecourt, architect of Cambrai cathedral. The album consists of thirty-three parchment leaves, with drawings of buildings, masonry, woodwork, geometrical figures, machines, human anatomy and animals. The embellishment of churches and chapels called for the services of the goldsmith, the colorist, the glazier and other craftsmen. Glass was a very ancient Egyptian discovery, but its widespread use for windows was a medieval development. Various technical treatises and manuals are extant: the *Compositiones ad tingenda musiva*, written in the time of Charlemagne, is a collection of recipes for preparing pigments, gilding materials, dyeing, and coloring glass; the *Mappae clavicula de efficiendo aura*, key to the recipe of making gold, is a more extensive manual of the same period. From the tenth century there has survived a book attributed to Heraclius, *De coloribus et artibus Romanorum*, and from the following century the more extensive *Diversarum artium schedula* of Theophilus Presbyter. This German monk, in his encyclopedia of arts and crafts, describes the preparation of lead-glass windows, the construction of organs, the casting of bells, and the making of oil colors, glues, varnishes, dyes, and inks.

Earlier mention has been made of medieval herbals and bestiaries and how their standard of illustration declined. But with the improvement of craftsmanship, this trend was reversed. The twelfth-century Bury St. Edmunds herbal has remarkably natural plant illustrations, clearly drawn from life. The Benedictine nun Hildegard of Bingen at the same time described a thousand plants and animals in her encyclopedia of natural history and "cloister medicine," the *Subtilitatum diversarumque creaturarum.*

We should also mention here the *De arte venandi* of the Norman emperor Frederick II. This book, completed before 1248, is devoted to falconry and describes the habits and structures of birds, the capture and training of birds of prey, and the practice of hunting with falcons. The margins of some existing manuscripts contain more than nine hundred drawings of individual birds, not only falcons, brilliant in coloring, accurate even to details of plumage. Frederick kept up a voluminous correspondence with his falconers and others, plying them with practical questions and instructions. He was deeply interested in animals—trailing with him around Italy and even across the Alps a menagerie that included elephants, camels, panthers, lions, leopards and the first giraffe recorded in Europe.

Frederick cited the zoological works of Aristotle, mainly to correct them. These works were unknown to Europe in the Dark Ages, and this is evidence that by this time new developments were taking place in the transmission of Greek science. It was, indeed, at Frederick's court in Sicily that the works had been translated, a decade or more before his own book was written. But they were far from being the first Greek works made known to western Europe: translations had been going on since the eleventh century, and it is this process that will now be considered.

CONTACTS WITH THE ARABS

The agricultural surplus painfully built up in Europe during the "dark" centuries led from the tenth century onwards to the gradual expansion of monastic settlements and other villages into towns. Indeed, Italian towns such as Genoa and Venice had begun to develop some centuries before. Over a thousand towns were founded in Europe during the Middle Ages. With town life came trade, breaking down parochialism and stimulating men's minds, and with trade came wealth and its concomitant leisure. Incentive and opportunity were provided to found universities and expand the tradition of learning. It is at this point that western Europe began to absorb the Arab culture that had preserved so much more thoroughly the science of antiquity.

Held to be the most learned scholar of the tenth century, Gerbert of Aurillac (Pope Sylvester II) visited Barcelona, and his later writings show some slight acquaintance with mathematical works that are Arab in origin. The earliest translation of an Arabic scientific text is said to be that of two treatises on astronomical instruments, made in the tenth century at a Benedictine monastery in Barcelona. John of Gorze—a monk from Lorraine—in the same century visited the monastery of Monte Cassino to acquire Greek manuscripts for his own monastery, and later lived for three years in Cordova, from which he may well have obtained Arabic manuscripts. Lorraine became the earliest center in western Europe for fostering the knowledge of Arab science. The eleventh-century astronomical texts of Herman Contractus, a monk of Reichenau, contain many Arabic words (and we have previously noted his gift to Chartres). Reichenau was one of the leading centers of learning in Germany—already in the ninth century its library contained treatises by Galen, Vitruvius, the astronomer Aratus and the physician Soranus.

The monastery of Monte Cassino and the town of Salerno (near Naples) played an especially important part in the early transmission of Arab and Greek science. The monastery (founded, as we have seen, by Benedict in 529) had long been famous for its manuscripts. There is extant in its library (or was, before its destruction by bombardment during World War II) a Galen manuscript of the ninth century and tenth-century copies of Dioscorides, as well as Hippocratic and other medical texts.

Salerno had been noted for medicine even in Roman times, and in the tenth century it was once more renowned for the practical skill of its doctors. In the middle of the eleventh century, Salerno was visited by an African merchant from Carthage. The story goes that his interest was so aroused that on his return to Carthage he studied medicine for several years, collected a number of Arabic manuscripts, and then went back to Italy. There he became a monk at Monte Cassino, took the name of Constantinus, and settled down to translating into or paraphrasing in Latin his precious medical texts—an optical treatise by Hunain ibn Ishaq, the Hippocratic *Aphorisms* and *Prognostics*, various writings by Galen, and other works. In due course, this material was assimilated by the physicians of Salerno, and became in the twelfth century the center of their medical curriculum. Salernitan commentaries were written on these texts, and their use spread to Naples and Paris. Substantially the same group of treatises was printed together in a popular medical collection, the *Articella*, first published in 1476: thus the translations of Constantinus formed one of the bases of medical teaching for several hundred years.

At the beginning of the twelfth century the pace of transmission of Arab and Greek science quickened. The conquest of Arab Spain by Christendom had begun: Toledo fell to the armies of Alfonso of Castile in 1085. Sicily, which had been under Arab domination since the ninth century, was conquered by Roger Guiscard of Normandy in 1091. The church Council of Clermont in 1095 decreed the necessity of "driving the infidels out of Palestine," and commissioned Peter the Hermit to lead a holy war—the first Crusade. The trade of Europe with North Africa and the East began to grow. All these events brought Western scholars into closer contact with the Arab world.

Most of the Crusaders were men of action rather than men of learning, but their relations with Islam were not wholly military. A Pisan, Stefano, followed his countrymen to Antioch in Syria, where in 1127 he translated into Latin an Arabic medical treatise (of which Constantinus had already given a partial version). To this Stefano added a glossary in Greek, Arabic and Latin of the technical terms found in Dioscorides.

Even more widely traveled was the Englishman Adelard. Born at Bath, at various times between 1109 and 1126 he lived in France, Salerno, Sicily, Syria and Palestine, and probably visited Spain. He translated from the Arabic the astronomical and trigonometrical tables of al-Khwarizmi (a ninth-century Persian) and fifteen books of Euclid's *Elements*. He also wrote his *Quaestiones naturalis*, professedly to explain to his nephew the new knowledge derived from the Arabs, discussing plants, animals, the psychology and physiology of man, meteorology and astronomy—a book that became very popular. Even greater Arab influence is shown in Adelard's later work on an astronomical instrument, the astrolabe. Through the efforts of this remarkable man, western Europe began to learn not only the facts and theories of Greco-Arab science, but also something of its spirit. The old world of Europe, Adelard tells his nephew, "follows a halter, caught by the appearance of authority, for what is authority but a halter? But I have learnt from the Arabs under the guidance of reason."

SICILY

The contacts between Europe and the Arabs established by such men as Constantinus, Stefano and Adelard were individual and intermittent. Systematic efforts to transmit Greek and Arab science to the West were soon to follow. The most important of these was centered on Toledo in Spain, as will be related in the next section, but attention must also be paid to the contribution of Sicily.

The cities of Sicily had been amongst the earliest Greek colonies. Syracuse was founded in 734 B.C. It was from the ruling family of this town that Archimedes was born in 287 B.C. From Agrigentum in the fifth century B.C. came the philosopher-physician Empedocles, founder of the Sicilian school of medicine. The capture of Syracuse by Rome in 212 B.C. brought Sicily into the empire. The emperor Augustus planted Roman colonies at Syracuse, Palermo and elsewhere, but Latin did not displace Greek—the two languages subsisted together. Germanic tribes ruled Sicily during the fifth century A.D., but in 551 the Eastern emperor Justinian drove them out and established relations between Sicily and Constantinople. This no doubt helped to maintain the tradition of Greek culture on the island.

In the eighth and ninth centuries, repeated Arab invasions took place, and the island was finally conquered in 878. Under two centuries of Arab rule, neither Christianity nor the Greek tongue died out, and the island was flourishing when captured by Roger Guiscard of Normandy. Tolerant rule continued—to the Greek, Arab and Hebrew languages already existing was added the French of the Normans. Settlers from Italy brought a new influx of Latin, at that time just differentiating into Italian.

Sicily in the twelfth century was thus a meeting-point of all the cultural elements of the Mediterranean area—Arabic, the vehicle of Greek science; classical Greek itself; and Latin, the international language of medieval Europe. Its Norman rulers had contacts with France and England, and the island maintained an active trade with Constantinople and Syria. In such a cosmopolitan atmosphere, the transmission of knowledge among cultures was inevitable. Translations were directly encouraged by the Norman rulers. It is worth noting also that close at hand on the southern mainland of Italy were a number of Greek monasteries, which had been founded at about this period by refugees from Constantinople, and in which Greek manuscripts were available.

The first Sicilian translations into Latin were in fact not from Arabic but directly from Greek. An officer of the Sicilian court, Henricus Aristippus, brought Greek manuscripts from Byzantium to Palermo, including a copy of Ptolemy's *Almagest*. This was translated by an unknown scholar in about 1160, aided by another court officer, Eugenius. Aristippus himself translated the fourth book of Aristotle's *Meteorology*. Other Sicilian scholars prepared Latin versions of Euclid's works on optics. This first phase of Sicilian transmission of science, from the Greek, was no doubt partly stimulated by the connection with Constantinople, which was undergoing something of a cultural revival at the time. This influence was temporarily ended in 1204, when Constantinople was seized and sacked by the Crusaders.

A second phase developed in the thirteenth century, during the reign of Frederick II. He established political and commercial relations with many Eastern and Western rulers and attracted many scholars to his court, the most famous being his court astrologer. Michael Scot—born in Scotland towards the end of the twelfth century—was also a link with the Spanish schools of translation. We first hear of him in Toledo in 1217, where he completed Latin versions of some of Aristotle's zoological works, all from the Arabic. He joined Frederick's court ten years later, and there translated the summary of Aristotle's zoology by Ibn Sina (Avicenna), and compiled a number of astrological and alchemical works.

SPAIN

The Spanish peninsula, wrote Haskins, is full of the visions of stories. "It has its romance of commerce, from the corded bales of the Tyrian trader to the silver fleets of the Indies; of discovery and conquest, as personified in Columbus and the conquistadors; of crusading and knight errantry in the Cid and Don Quixote. It also has its romance of scholarship. . . . In consequence of the Saracen conquest, the peninsula became a part of the Mohammedan East, heir to its learning and its science, to its magic and astrology, and the principal means of their introduction into western Europe. . . . In the twelfth and thirteenth centuries, the great adventure of the European scholar lay in Spain."

The country did not possess any elements of Greek culture lingering from classical times, as did Sicily, so that direct transmission of science from the Greek played no part. Spain was, however, a meeting place of three linguistic groups—Arabic, Latin and Hebrew. Down to the twelfth century, contacts between Christian and Islamic Spain seem to have been small, but not long after the fall of Toledo to Alfonso in 1085 translations began.

Although the work went on in several cities of northern Spain, Toledo was of most importance. At this ancient center of scientific teaching were to be found a wealth of Arabic books and men who spoke both Arabic and Latin. Something approaching a translator's academy was organized by the archbishop of the city in 1126. Two of the earliest members of the school were Ioannes Hispalensis (John of Seville, Avendeut) and Dominicus Gundisalvus (Gundissalinus). Ioannes translated from Arabic into Castilian, and Dominicus then produced a Latin version. Between them they translated over twenty books from the Arabic, including *De scientiis* by al-Farabi (from the tenth century), an astronomical manual by al-Fargani, and an elaboration of al-Khwarizmi's arithmetic. Dominicus himself wrote a book on the classification of science, much influenced by al-Farabi.

The most industrious and prolific of all the translators who worked in Spain was Gerard of Cremona (1114–87). He was drawn to Toledo in search of Ptolemy's *Almagest*, of which he had heard but which he could not find in Latin (the Sicilian translation of 1160 had probably not then been made, and anyway Gerard did not learn of it till later). At Toledo, he found a multitude of Arabic books on every subject and decided to learn Arabic and devote his life to translating them. He is credited with the production of nearly ninety translations—a number that strongly suggests he was the head of a group of workers, as Hunain ibn Ishaq had been in Baghdad three centuries earlier.

Gerard's translations ranged over philosophy, mathematics, astronomy, physics, medicine, astrology and alchemy. Some of the most important are logical and philosophical texts by Aristotle and Arab writers; Aristotle's books *On the Heavens*, *On Generation and Corruption*, and *Meteorology*, and commentaries on these by Alexander of Aphrodisias; the *Elements* of Euclid and the *Conics* of Apollonius; the *Almagest* of Ptolemy and other works of mathematical astronomy; Galenic texts; the *Canon* of Ibn Sina (Avicenna) and the *Continens* of al-Razi (Rhazes).

Translations were not made only at Toledo. At Barcelona, the Italian Plato of Tivoli between 1135 and 1145 translated a number of Arabic works such as al-Battani's *Motions of the Stars*. Some of his translations introduced Arab trigonometry and mensuration to the West. Herman Contractus in 1143 provided a translation of a work by Ptolemy, the *Planisphaerium*, that is not available in the original Greek. Robert of Chester collaborated with Herman in translating the Koran, but his real interest lay in science. He made the first Latin version of al-Khwarizmi's *Algebra*, and was responsible for one of the earliest translations from Arabic to Latin of an alchemical text.

TRANSMISSION INTO LATIN COMPLETED

The great task of translating scientific works from the Arabic into Latin was largely completed during the twelfth century. Translation still continued—also from Arabic into Hebrew—and there will be occasion to mention a few more scholars who contributed to this work, but the thirteenth century was more concerned with expounding its newly acquired learning in original but derivative works.

Right at the beginning of the century stands a scholar who combined the two activities, the Englishman Alfred of Sareshel. He is known to have visited Spain, and also appears to have had knowledge of Greek texts of the works he translated. The most important of these were the pseudo-Aristotelian treatise *De plantis*, and a book on minerals by Ibn Sina, which was also long considered to be by Aristotle. Alfred wrote commentaries on other works by Aristotle, and opens yet another phase—an extension of the twelfth-century concentration on mathematical and astronomical writings to an interest in natural history which developed in the thirteenth century.

Several important works expounding Greco-Arab mathematics and astronomy were written at this time. Leonardo of Pisa (Fibonacci) lived for some years at Bugia on the Barbary Coast, where he learned Arab mathematics. During the period 1200–25, after his return to Pisa, he wrote several books on this subject: the *Liber abacus*, *Practica geometrica*, and *Liber quadratum*, all three important in the development of European mathematics. A little later, Arabic astronomy was influentially expounded in the *Sphaera mundi* of John of Holywood (Sacrobosco). In about 1272, Jewish scholars gathered by Alfonso el Sabio at his court in Toledo completed in Castilian some new astronomical tables, translated into Latin by John of Saxony in the fourteenth century.

Medical science also had further translators and expositors. In about 1255, Bonacosa of Padua translated the *Colliget* of Ibn Rushd (Averroes), and later Simon of Genoa compiled a pharmacological dictionary of six thousand terms, drawn from the works of Celsus, Dioscorides, Pliny, Galen, al-Razi, Ibn Sina, Constantinus and others—a clear indication of the wide knowledge of Greek and Arab writings that had by then been obtained. Arnaldus Villanovanus—to whom many medical and alchemical texts are attributed—also translated works by Galen, Ibn Sina and others from the Arabic.

Despite the prolonged efforts of the translators, the knowledge of Greek science achieved by western Europe was still far from complete. Many authors were still unknown. Moreover such texts as were available were often far from accurate. The Latin versions were sometimes fourth hand, being derived from the original Greek via a succession of Syriac, Arabic and even Hebrew intermediates. Comparison with such Greek texts as were preserved in Sicily or Italy soon led scholars to feel the need to find new Greek sources from which they could translate. The work of William of Moerbeke (1215–86), bishop of Corinth, is the outstanding example of this new development. He made Latin translations directly from Greek manuscripts, most probably obtained from Constantinople, which had been captured by the Crusaders in 1204. The texts he worked on included various Galenic and Hippocratic treatises, most of the works of Archimedes that were then available, Hero's *Catoptrica*, Aristotelian commentaries by Alexander of Aphrodisias and Simplicius, and almost the complete works of Aristotle.

An interesting reflection of the route whereby Greek science came to western Europe was given in the *Tabulae* of the Renaissance anatomist Andreas Vesalius (1538). His illustrations indicated the name of each part of the body in Latin, Greek, Hebrew and transliterated Arabic.

ENCYCLOPEDISTS AND COMMENTATORS

Just as the Arabic translations of the ninth century were rapidly followed by original works expounding and interpreting the new knowledge obtained from the Greeks, so too in thirteenth-century Europe classical and Arab science was popularized and discussed in compendia and commentaries. Indeed, examples of this occurred even earlier. We have already mentioned the work of Adelard, and in the twelfth century, Alexander Neckam of St. Albans wrote a work of instruction, *De natura rerum*, that included much popular science.

During the twelfth and thirteenth centuries in Europe, as we have seen, there was a great development of town life. As trade and manufacture increased, knowledge of reading and writing became more widespread, and the desire for education grew. The old educational syllabus based on the seven "liberal arts" was clearly incapable of accommodating the wealth of new knowledge that the translators had revealed. On the other hand, the great volume of writings now available could not readily be assimilated by a student. Some simpler sources of instruction were needed.

The task of popularizing and spreading the new knowledge was taken up by the orders of friars—Dominicans and Franciscans—that were founded early in the thirteenth century. They tackled the problem by writing systematically arranged encyclopedias, in which were collected the opinions of the best-known authorities. These works became the textbooks and reference books of teachers and students. The medieval encyclopedias thus had a clear educational aim. As one of the most popular authors put it, "I write for students and for those of moderate culture who, being unable to refer to the innumerable works which treat of the properties of things mentioned in Holy Scripture, cannot readily find even elementary ideas concerning them" (Bartholomew the Englishman).

The Dominican Thomas of Cantimpré, in the period 1228–44, compiled *De natura rerum*. "With labour, with immense care," he wrote, "I have worked for fifteen years, reading the works of many philosophers, in order to bring together in one small book all that I have found useful and edifying on the nature and properties of created things." More than half his text was about animals, though he also dealt with plants, waters, stones, metals, stars and man.

The Franciscan Bartholomew in 1230–40 compiled a very popular encyclopedia, *De proprietatibus rerum*, written as an educational text for the common people ("simplices et rudes"). His work, like all the medieval encyclopedias, is a compilation of extracts from many classical authorities—Aristotle and his Arab commentators, Constantinus, Galen, Hippocrates, Ibn Sina, Ptolemy and so on, as well as the old standard texts of Pliny, Dioscorides, Isidore and Bede. Each extract is followed by a comment by Bartholomew himself. He thus discusses God, angels, the mind and body of man, domestic economy, medicine, cosmology, geography, animals, plants, minerals and much else. His text, translated into English by John of Trevisa in 1397–98, was one of the earliest specimens of English prose, and this translation was the first scientific work of any importance to be printed in English, being published at Westminster in 1495 by Wynkyn de Worde.

Vincent de Beauvais, librarian and tutor to Louis IX of France, produced an encyclopedia greater in size than any other of his time, the *Speculum maius*. It included an account of natural science in the form of a gigantic commentary on the first chapter of Genesis; a summary of theoretical knowledge (such as mathematics) and practical knowledge (such as medicine, agriculture, architecture, navigation, metallurgy, etc.); a universal history; and even a dictionary of three thousand words. During 1260–67, Brunetto Latini of Florence while in France wrote *Li livres dou tresor*, a popular encyclopedia mainly devoted to geography and zoology.

Brief consideration must be given to three scholars who dominated the thirteenth century. They passed beyond the compilation of extracts, and their work heralds the new experimental science of the Renaissance. The Dominican Albertus Magnus produced a series of large commentaries on Aristotle, in which he incorporated the results of much keen personal observation. Robert Grosseteste, bishop of Lincoln from 1235 to 1253 and first chancellor of the university of Oxford, is held by some to be the founder of scientific thought in England. He wrote works on sound, heat, color, geometry, comets, optics, the astrolabe, poisons and other topics.

The most famous writer of the century was yet another Englishman, a Somerset man, the Franciscan Roger Bacon, who studied at Oxford under Grosseteste. Roger Bacon had an intimate knowledge of all the scientific writings available to his age, and was himself a practical experimenter. Like Adelard before him, Bacon disparaged the authorities of antiquity—"we find their books full of doubts, obscurities and perplexities"—advocating both the use of reason (in the shape of mathematics) and experiment. He highly praised his contemporary Peter Peregrinus, who experimented with the magnet. Bacon wrote of him:

> he knows natural science by experiment and medicaments and alchemy and all things in the heavens or beneath them, and he would be ashamed if any layman or old woman or rustic or soldier should know anything about the soil that he was ignorant of. Whence he is conversant with the casting of metals and the working of gold, silver and other metals and all minerals; he knows all about soldiering and arms and hunting; he has examined agriculture and land surveying and farming. . . .

"Never was there so great an appearance of wisdom, nor so much exercise of study in so many faculties, in so many regions," wrote Bacon of his age. "Doctors are dispersed everywhere, in every castle, in every burgh, and especially among the students of the two religious orders." Western Europe had received the full heritage of Greek and Arab science, and was slowly beginning to build anew on this foundation. New ideas were being considered everywhere, though never very clearly. Discoveries were made and lost again, because the inadequacy of communication often kept manuscripts virtually unknown outside a small circle. Nevertheless, the first stirrings of modern science were occurring.

MEDIEVAL UNIVERSITIES

From the cathedral schools mentioned earlier arose one of the Middle Ages' greatest contributions to our modern civilization—the university. The later universities had definite foundations to which dates can be assigned, but this is not possible with the earliest, which gradually came into being among the assemblies of students and masters attracted to one or another of the larger cathedral towns by the fame of its school or teachers. In Italy and Provence a vigorous town life was in existence by the twelfth century. These were the regions where Roman influence had been strongest, and the

Lombards and Burgundians had adopted many Roman customs and institutions. Roman law remained in force alongside the laws of the conquerors. The judge, the advocate, the notary and the municipal administrator became indispensable members of society.

Italian schools began therefore to teach law at an early date. It is claimed that at Pavia, the capital of Lombardy, there was a vigorous law school as early as the eighth century. No wonder then that the university that arose in Bologna, one of the most favorably placed towns in northern Italy, should have early achieved a reputation as a center for teaching law, second to none in Europe. The university appears to have grown out of three different systems: the arts school attached to the cathedral, municipal schools that taught Roman civil law, and monastic schools teaching church law. It is in 1158 that we first find in Bologna mention of recognition of the masters and students as corporate bodies. The students had organized themselves into a guild, and the university took the form of a corporation of students. The masters formed themselves into a "collegium doctorum" to supervise admission into their ranks and to regulate their own affairs.

In France, matters took a somewhat different course. There, Roman town life after the collapse of the Western empire had rapidly given way to a feudal economy based on the countryside. The illiterate rural nobility relied on the church for its administrators. The church schools of France were particularly noted for their teaching of theology and its necessary preparation, the seven liberal arts, and this bias was to be continued in the new universities. In 1100, there were three schools in Paris—those of the cathedral of Notre Dame, of the canons of St. Victor, and of the church of St. Genevieve. During the ensuing century, the university gradually took shape, essentially in the form of a corporation of masters.

The word "universitas" originally meant no more than a corporate body of some kind, and medieval universities at first consisted entirely of the guilds of masters and/or students. If these harbored a grievance against the city—and disputes between town and gown were frequent—they could pack up their belongings and depart in a body for a more friendly town. Secessions from Bologna, for example, were the occasion of the founding of several other universities. Oxford University was formed during the twelfth century, once again on the basis of existing church schools. A secession from Paris in 1167 brought it added strength, while a secession of scholars from Oxford in 1209 led to the formation of a university at Cambridge.

Two towns that played a great role in the transmission of knowledge should be mentioned: Salerno and Montpellier. They did not fit into the pattern described above, for they were renowned not for cathedral schools but for medical ones. Salerno, as we have seen, had long been noted for its medical men. Legend has it that the medieval "school of Salerno" was founded by a Western Christian, a Greek, a Jew and an Arab, and the story (fanciful as it may be) indicates the influences that combined to make Salerno the foremost center of medical teaching in Europe. At first there was no regular school—students attached themselves to individual masters. But in the twelfth century, would-be physicians were flocking to Salerno from all over Europe, and a loose association of masters and students was formed.

Montpellier in Provence had considerable contacts with Italy and Spain. The first notice of a medical school there is in 1137, and by 1160 the school had gained a reputation second only to Salerno. The formation of a university took place over the period 1180–1220: the masters formed a legal body with the right of examining and granting licenses to teach. Medicine was one distinct faculty, the others being law, theology and arts. Montpellier retained its position as a great study center until the middle of the fourteenth century and its reputation as a medical center for some time after. It was at Montpellier that François Rabelais in the sixteenth century announced a remarkable innovation—lectures on the original Greek text of Hippocrates.

The late twelfth and thirteenth centuries saw the beginning of a number of other universities taking Bologna or Paris as their model. In Italy, Pisa, Reggio, Vicenza, Arezzo, Padua, Florence and Siena; in France, Orleans, Angers, Toulouse; in Spain, Salamanca, Valladolid, Seville; Coimbra and Lisbon in Portugal. As an indication of the ultimate spread of the university idea, appendix A lists European and other university foundations up to 1900.

The formation of universities expanded the number of students and began to provide a means of supporting scholars who could contribute to the intellectual development of Europe. From this time on, scholars (and eventually scientists) could teach and research within the framework of an official institution, with the possibility of a guaranteed salary, working accommodations and other facilities. We should note, however, that this did not necessarily mean a settled life in one place. Mention has already been made of secessions of scholars from one town to another. But apart from this, students were restless. What knowledge they could not find at one university, they sought at another. Travel was becoming easier—more transport, wayside inns—and even those without much wealth could undertake a "study tour," travelling from master to master. The universities in turn began to compete, to seek out the most eminent teachers, whose presentations would entice and retain students.

BOOKS AND LIBRARIES

In addition to the masters and students, the medieval university attracted to itself a motley collection of clients who catered for the many needs of the establishment without actually being members of it. As time went on the university assumed greater control over them, but in compensation for some loss of independence they enjoyed the monopoly of trade in the university, in whose precincts they were allowed to set up business. Among these, the *stationarii* were of particular importance.

The *stationarius* combined several functions that are today shared between several different people: printer, publisher, bookseller, circulating librarian and stationer in the modern sense of the term. Besides providing books he also sold to masters and students their pens and parchment and other writing equipment. He employed scriveners to prepare copies of manuscripts and might himself take a hand at copying. Since so many students could not afford to buy a book outright, he also hired out manuscripts for copying by the borrower. To ensure that copies reached an acceptable standard of accuracy, the university regulated the stationer's trade—only certain stationers were licensed to supply books to students, subject to special ordinances.

Book production in the university was based on the exemplar and the *pecia*. The exemplar was a manuscript that had been thoroughly scrutinized by a commission of scholars appointed by the university and declared by them to be an authoritative text. It formed a master copy from which others were derived. It was kept in loose sections, *peciae*, and scriveners were given one *pecia* at a time, so that several scribes could be engaged simultaneously in copying a work. The *pecia* was likewise the unit for lending.

In 1308 at Oxford, a stationer is mentioned by name, one Robert, although from other references it is clear that copyists were active long before then. By 1374 there were so many booksellers in Oxford that the university decreed that the trade in books valued at more than half a mark should be restricted to sworn stationers. In time the book trade became highly centralized: by the fifteenth century there was only one university stationer at Oxford.

It should be mentioned that the development of this new kind of book trade coincided with the gradual introduction of paper as an alternative to parchment. Paper manufacture, originating in China, had been going on in Persia and Syria since the eighth century and had spread to Spain by the twelfth. Pacey has given a picture of the diffusion of papermaking, reproduced in table 2. Universities such as Paris and Caen in the fourteenth century authorized the construction of mills to supply paper for their needs.

For many years the universities did not have libraries as such. Each teacher would have a small collection of his own books, which could sometimes be lent to students for copying. Probably the earliest "libraries" associated with the university, other than the stocks of the stationers, were those of the student groups or "nations." These groups sometimes lived together or at least had accommodations where books were communally owned and used. At Oxford and Cambridge they gradually developed into "colleges," each with its own faculty, curriculum, and later its own library. Although many of the colleges making up the university date from the thirteenth century, their libraries seem to have begun in the following century or even later.

As an example of the slow development of a general university library we may continue to look at Oxford. Johnson notes that about 1320 Thomas

Table 2 Diffusion of Papermaking

Place	*Date (A.D.)*
China	100
Tibet	650
India	670
Samarkand	750
Baghdad	794
Cairo	850
Damascus	1000
Tripoli	1000
Sicily	1000
Fez	1050
Jafiva	1150
Montpellier	1250
Fabriani	1276
Nuremberg	1390
London	1490

Source: Pacey (1990)

Cobham, bishop of Worcester, erected a building at Oxford, part of which was to be used as a library, but the project was not completed, and his books ended up in Oriel College library. The small collection drifted until 1411, when a new chancellor of the university, Richard Courtenay, took it under his supervision. The collection grew, with many gifts from Humfrey, Duke of Gloucester, and others, and in 1480 it moved to a new location (the Divinity School building), where it became known as "Duke Humfrey's Library" in honor of its chief donor.

DEVELOPMENT OF EUROPEAN LANGUAGES

Intermittently in this book, mention has been made of writings in the vernacular languages of Europe. It is now time to look at their development in more detail. The modern languages of Europe are the result of a complex mingling of cultural groups over several millennia. They are all (except Basque, Finnish and Magyar) members of the Indo-European group of languages that includes Celtic, Greek, Latin and the Slavonic and Germanic tongues. The speakers of these languages spread from their original homeland in eastern Europe, and by about 600 B.C. we find the Latins in Italy, the Greeks around the Aegean, the Celts or Gauls occupying Europe from the lower Danube to Britain, the Germanic tongues to the north and the Slavonic to the east.

The conquests of Rome imposed the Latin language upon the inhabitants of France and Spain, but this was not effected without changes in Latin itself. In each region of the empire it was partly corrupted into a local dialect. There was even a vulgar dialect—*rusticus*—in Rome itself. The Germanic invaders of the fourth and fifth centuries—Vandals, Goths, Franks and others—though they ultimately adopted these neo-Latin dialects, altered them still further by the introduction of Germanic words, idioms, phrases and constructions. French thus began as a provincial Latin dialect as modified by the Germanic-speaking Franks.

The use of classical Latin continued after the invasions, particularly in church circles, though steadily declining from the standards of classical literature. The clergy of France still preached in Latin during the eighth century, but at the Council of Tours in 813 it was ordered that homilies should be explained to the people in their own tongue, whether rustic Roman or Frankish. By the eleventh century, two distinct forms of French were in existence—the northern dialect and Provençal. In Italy, as might be expected, the corruption of Latin was less extensive and less rapid, and there is little evidence of an independent language before the twelfth century.

The linguistic development of regions that had not been fully romanized was different. They remained relatively free from Latin. An early example from the Germanic group of languages is the fourth-century translation of the Bible by Ulfilas, which was used by all the tribes who later advanced into Spain and Italy. Other Germanic dialects gave rise to the English, Dutch, German and Scandinavian languages. In England, the language of the fifth-century Saxon invaders was successively modified by the impact of further invaders, the Danes and the Normans, and became recognizably English in the period 1100–1250.

The development of the vernaculars into literary and administrative languages was a slow process. Latin remained the accepted medium for the arts, sciences and government throughout western Europe; Greek in what was left of the Roman empire in the East; and Old Church Slavonic in Russia and the rest of the Slav world. For the mass of the peoples Latin became an obscure and learned tongue. The Irish monk or the Bavarian priest, who spoke Old Irish or Early German, had to learn church Latin for his vocation. To learn the more elaborate classical Latin was still more difficult and confusing. Few churchmen did so—particularly as there was always opposition within the church to any study of the writings of the corrupt pagan civilization.

It is little wonder that in these circumstances the level of literacy up to A.D. 1000 was very low indeed. The language of culture and the language of common speech never coincide—even in modern English, in which the literary and spoken forms have come closer together, the coincidence is by no means complete. In feudal Europe, with a more or less international aristocracy opposed to national peasantries, the gulf was particularly wide, and was made more difficult to bridge by the multiplicity of regional dialects and the paucity of vocabulary in any of them. Any idea of more than regional significance was consequently compelled to clothe itself in Latin. This could only change when a regional dialect could win national acceptance and widen its vocabulary. Vernaculars grew first into literary languages, and only later challenged Latin for the communication of philosophical and scientific notions. A few examples of this development will be noted.

The most important early Anglo-Saxon poem, *Beowulf*, survives in a manuscript of about the year 1000, but English was certainly written some centuries earlier than this. In the eighth century, Bede translated some of the New Testament into English; in the ninth, King Alfred participated in the translation from Latin to English of the *Consolations of Philosophy*, written by Boethius. The main prose work before the Norman conquest of Britain was the *Anglo-Saxon Chronicle*, composed during the tenth and eleventh centuries in various monasteries.

It is considered that Latin had developed into French by about the eighth century. The earliest literary texts are from the tenth century, and the first known major work in French is the eleventh-century *Chanson de Roland*, the first of a large number of heroic poems dealing with adventure and war in the Western world, with heroes such as Charlemagne, Alexander, the Trojans or King Arthur and his knights. The oldest document in Italian is from the twelfth century, and at the end of the thirteenth century Dante's *Divina commedia* fixed the Florentine dialect as the literary language. In Spain, we have the twelfth-century *Song of El Cid*.

The first use of the vernaculars of western Europe for scientific writing came about in the thirteenth century, particularly in French. In the following century, the vernaculars became a serious challenge to Latin in the more popular writings. Sarton gives the results of counting scholarly authors in various tongues in table 3.

For the whole fourteenth century, more than a third of western European scholarly authors wrote in the vernacular, and the percentage increased from 27 percent in the first half of the century to 42 percent in the second half.

The need for specialized vernacular texts was felt first in the more practical fields, such as medicine, agriculture, military engineering and computation, as well as among the popular travel books and encyclopedias. The first known technical author in German was Conrad of Megenburg (although Hildegard of Bingen in the twelfth century had quoted the German names of a thousand animals and plants in her *Physica*). Conrad in 1340–

Table 3 Latin and Vernacular Authors

Language	*1300–1350*	*1350–1400*	*1300–1400*
Latin	223	192	415
French	22	51	73
Italian	24	30	54
German	12	15	27
English	4	10	14
total non-Latin	82	137	219

Source: Sarton (1938)

50 wrote *Das Buch der Natur*, and made a German translation of the thirteenth century John of Holywood's astronomical work.

A "leech-book" in Anglo-Saxon dates from the tenth century, describing five hundred plants and their medicinal uses. The Norman conquest of England powerfully influenced the development of the language, which by the thirteenth century was more recognizably akin to modern English, but the conquest also had the effect of temporarily stifling the use of English as a literary language. The court spoke French, scholars wrote in Latin. Apart from manuals such as Walter of Henley's *Hosebondrie*, English prose did not develop until the middle of the fourteenth century. The mythical *Travels* ascribed to John Mandeville were then translated from the French. Chaucer translated the *Roman de la rose*, and as well as his *Canterbury Tales* wrote two technical works on astronomical instruments, the *Tretis on the Astrolabe* and the *Equatorie of the Planetis*. John of Trevisa put into English the *Properties of Things* of Bartholomew. The earliest English medical text was a 1380 translation from the Latin of Lanfranchi's *Chirurgia magna* (written in 1296).

The first technical texts in French occur in the thirteenth century. There is an anonymous French "algorisme" of this era, and Walter's *Hosebondrie* was first written in French (in about 1250). In 1256–57 the Italian Aldobrandino of Siena compiled a French medical treatise, and a few years later Brunetto Latini, another Italian, produced an encyclopedia, *Li livres dou tresor*. In 1298–99, Marco Polo's travels were first narrated by Rusticello in French.

In the fourteenth century, a number of medical works appeared in the French language, such as translations of the aphorisms of Hippocrates. Jean Corbechon translated Bartholomew's encyclopedia and a Latin agricultural text by Pietro dei Crescenzi. Nicole Oresme—one of the most original scientists of the century—produced the earliest vernacular versions of Aristotle, including his work *On the Heavens*, and wrote a treatise on mathematical geography. All the works so far mentioned were in the northern dialect of France. The troubadours had written their lyrical poems in the southern dialect, Provençal, and this too produced a flood of technical texts—mostly translations of medical works, algorisms, bestiaries, treatises on alchemy, and the universally popular Bartholomew.

Alfonso X, king of Castile 1252–84, organized the translation from the Arabic into Spanish of a large number of astronomical works by Ptolemy and Arab writers. A major Spanish text, the *Libros del saber de astronomia*, was primarily an encyclopedic account of astronomical instruments. Brunetto Latini translated some of Aristotle into Italian in 1290, but the earliest technical use of the language was for texts on commercial arithmetic, needed by the bankers and traders of Italian cities. Texts by Francesco Pegolotti (1340) and Paolo Dagomari (1339) were the first of a long line of notable manuals. The Italian text of Marco Polo's travels, *Della maravigliose cose del mundo*, was established before his death in 1324. In the same period, Pietro dei Crescenzi's agricultural text was put into Italian.

An interesting byway of linguistic history is the view expressed by Dante, early in the fourteenth century, on the legendary confusion of languages arising from the construction of the Tower of Babel. In this construction,

> almost the whole human race took part: some gave orders, others made sketches, others erected walls, some aligned them, some put up plaster, others crushed rocks, some transported them by sea, others by land, and some were occupied in all kinds of other work. So that when the blow from the heavens was delivered into such a mixture, all who had spoken at work in one and the same language, now spoke a multitude of completely different languages. They ceased work and most of them could not agree on anything. Indeed, only among those occupied with the same task was retained the use of the same language. For example, there was one language for architects, one for all transporters of rocks, one for all rock crushers; and so it happened with all the specialties of the workers; and for as many different separate occupations of intellectual work there were, there was an equal amount of languages, and from that time the human race was so divided.

Dante's observation offers a comment on professional jargon, and on the separation of craft and intellectual knowledge.

During the thirteenth and fourteenth centuries, all the leading languages of western Europe developed into instruments flexible enough to expound a wide variety of scientific and technical matter. Nevertheless, scholarly writers were slow to take up the vernaculars, understandably enough, for by writing in Latin a scholar could command an international audience greater than that of any national language. Latin, too, still gave a certain cachet to a learned work, and a long time was to elapse before it ceased to be the major language of scholarship.

HUMANISM

A final stage in the recovery of the classical heritage was the work of the humanist movement, centered in fourteenth- and fifteenth-century Italy, spreading later to other European countries. Among its distinctive features were an emphasis on Greek texts rather than Latin, on Plato rather than Aristotle, and on the revival of the library of the wealthy or subsidized private scholar—a type of book collection that had almost disappeared since the decay of Roman villas.

The birth of humanism is often associated with the poet Petrarch. Born near Avignon (his family were exiles from Florence) he early discovered the monastic and cathedral libraries of France within reach, and began a lifelong hunt for classical manuscripts. He wrote in 1346: "I am still in the thrall of one insatiable desire which hitherto I have been neither able nor willing to check. I cannot get enough books. It may be that I have already more than I need, but it is with books as it is with other things: success in acquisition spurs the desire to get still more." He built up a classical library unrivalled in the fourteenth century, and he read and reread its contents. Petrarch discovered books lying forgotten in ecclesiastical collections and had them copied and distributed to his friends, that they might be read and discussed. In later years he tried to bequeath his books to the city of Venice, for use in perpetuity "by all scholars and gentlemen of the city," but his plans came to nothing. Boccaccio, writer of the *Decameron*, followed Petrarch's enthusiastic example, and gave up writing in Italian to search for Latin and Greek texts, and to write encyclopedic texts on classical biography, geography and mythology.

In the fifteenth century the humanist search for manuscripts continued. For example, Poggio Bracciolini of Florence made expeditions to libraries in Cluny, St.Gall, Fulda and Cologne, and spent some years in England. Soon the bulk of Latin literature now known to us had been unearthed from obscurity. One of Poggio's discoveries was the Latin poem of Lucretius, *De rerum naturae*. Meanwhile, the active collection of Greek texts from Constantinople and other Eastern cities had begun. Greek studies had been successfully inaugurated in Florence by Chrysoloras, a visitor from Constantinople, about 1400. A wealth of Greek manuscripts was brought to Italy by collectors such as Aurispa and Filelfo early in the fifteenth century. Guarino of Verona discovered a scientific text that had been lost for centuries, the encyclopedic medical treatise of the first-century Celsus.

Libraries were built up—for example, by the cardinal Bessarion, a Greek resident in Italy, his collection eventually becoming the Biblioteca Marciana in Venice; by the duke of Urbino; by the Medici family in Florence, at the convent of San Marco and the Biblioteca Laurentiana; and in the Vatican. In 1462, Cosimo de Medici set Marsilio Ficino, the son of his court doctor, the task of translating into Latin the complete works of Plato. Around Ficino there grew up a circle of scholars that has been called the "Platonic

Academy of Florence," though it hardly merits such a formal name. Though humanism had relatively little interest in or contact with scientific and technical texts, it created a new tradition of secular scholarship, shifting learning away from its medieval monopolization by the church. The support of humanist scholars by wealthy rulers would be followed in due course by their support for scientists.

LITERACY AND BOOK PRODUCTION

By the fourteenth century, a new reading public was being formed in Europe. As well as clerics, nobles and a handful of scholars, there were lawyers, lay counselors of kings, administrators of all kinds, rich merchants, master craftsmen. All had a need for books dealing with their specialty (such as law, accountancy or technology), and vernacular literary works were also eagerly sought.

Growth of an urban society and development of schools and literacy were closely related phenomena. Up to the fourteenth century, Italy was in the forefront of urban development. Moreover, lay schooling had, although much weakened, managed to survive there throughout the Dark Ages. It is reported that at the end of the thirteenth century Milan, with a population of sixty thousand, had more than seventy teachers of elementary schools. In the fourteenth century it was claimed that in Florence around ten thousand boys aged from five to fourteen were at school. This exceptional educational effort must obviously be related to the equally exceptional cultural flowering in Florence at that time. But this was not typical of the rest of Europe.

England, for example, was a very much stratified society. As previously noted, the upper classes and most educated persons spoke French, clerics and scholars used Latin for intellectual and educational purposes, peasants and townsmen spoke English dialects. Court records and other public documents were in either Latin or French. Langland and Chaucer were beginning to create English poetry. Even among the clerics (thirty thousand ordained clergy, fifteen thousand monks, seven thousand nuns) literacy was very unevenly spread. On the other hand, there were growing numbers of literate laymen—merchants and moneylenders in such towns as London, York, Bristol and Norwich; lawyers; skilled craftsmen (particularly the master masons). Overall, it has been estimated that in a total population of three million, perhaps 3 percent were literate.

As the French influence in England was shaken off, and new grammar schools were founded (e.g., Winchester in 1382, Eton in 1440), literacy increased among the nobility and gentry. The correspondence of the Paston family (from the fifteenth century) reveals a remarkably high level of vernacular literacy, not only within the family but also among their agents and bailiffs. In 1422, the London Brewer's Company resolved to keep their records in English because, they reported, many members could read and write it, "but Latin and French they do not in any way understand."

Because of the greatly increased demand for reading matter, there developed workshops of copyists producing manuscripts for sale by booksellers in a thoroughly commercial fashion. A possible London establishment dated 1330–40 has been suggested. An example of a text that must have been produced under some such conditions is the *Travels* of John de Mandeville, of which 250 manuscripts have been preserved (73 in German and Dutch, 37 in French, 40 in English, 50 in Latin, plus Spanish, Italian, Danish, Czech and Irish versions).

In Italy, writes Hessel,

> the most famous of the commercial dealers was Vespasiano da Bisticci, whose shop served as a rendezvous for the literary world. He had broad bibliographical knowledge and assembled entire libraries from near and far for his noble customers. By using 45 scribes he once delivered on order to Cosimo de Medici 200 volumes in 22 months. . . . One of his best customers was the great bibliophile, Federigo of Montefeltro, duke of Urbino. This nobleman appears to have spent greatly on his collection and to have kept steadily occupied 30 to 40 copyists. Place was provided for the luxurious and splendidly ornamented manuscripts in rooms specially set aside for this purpose in his newly erected castle.

Also in response to the demand for cheap texts, there was a steady increase in the use of paper (less expensive than parchment), as we have noted in the case of university stationers. It is clear that the time was ripe for a new mechanism for the mass reproduction of texts.

ORIGIN AND SPREAD OF PRINTING

In its promise and subsequent achievement, the development of printing looks forward to the modern world, but it is convenient to treat it here as a final phase in the transmission of recovered classical texts to a wider public. Apart from the classics and of course the Bible, relatively little that was printed during the fifteenth century has continued as living literature or scholarship.

The techniques that revolutionized fifteenth-century book production had antecedents stretching a long way back in time and space. Paper and printing both began in the Far East. Earlier we have looked at the westward diffusion from China of paper, which came fully into its own in Europe—completely displacing parchment—only with the advent of printing, the history of which also begins in China. Block printing probably started there in the sixth century. Writing many centuries later, in 1318, the Persian al-Banakati left an account of what he had learned of Eastern printing:

> The Chinese were wont and still continue to make copies of their books in such wise that no change or alteration can find its way into the text. . . . They order a skillful calligraphist to copy a page of the book on a tablet in a fair hand. Then all the men of learning carefully correct it, and inscribe their names on the back of the tablet. Then skilled and expert engravers are ordered to cut out the letters. When they have thus taken a copy of all the pages of the book, numbering all the blocks consecutively, they place these tablets in sealed bags. . . . Then when anyone wants a copy of the book he pays the dues and charges fixed by the government. Then they bring out these tablets, impose them on leaves of paper, and deliver the sheets to him.

The earliest book extant, printed in this way, is dated 11 May 868.

However, the Chinese advanced beyond block printing, and an eleventh-century writer, Shen Kua, has left an account of the invention of movable type. During the period 1041–49, a craftsman Pi Sheng cut characters as thin as the edge of a coin in clay, each character forming a single type; he then baked them in the fire to make them hard. He had previously prepared an iron plate and had covered it with a mixture of pine resin, wax and paper ashes. When he wished to print, he took an iron frame and set it on the plate. In this frame he placed type, set close together. When the frame was full, the whole made one solid block of type, and this was heated to bed the characters in the wax base, the whole being smoothed to an even height by pressure from a flat board. The type was inked for imposition on paper. "For printing only two or three copies this method would be neither convenient nor quick," wrote Shen Kua, "but for printing hundreds or thousands of copies it was marvellously quick." In 1314, Wang Chen printed his work on agriculture with movable type made of wood. By 1403, in Korea, bronze had been adopted for this purpose.

In Europe, the block printing of woodcut illustrations goes back to the fourteenth century. Block books—with both text and pictures—date from about 1430 and were almost wholly biblical in content. A few years after this, experiments in typography began. The development of the manuscript trade and the distribution of block books are evidence of the growing demand for books; paper was available as a cheap and abundant material for taking inked impressions; metallurgy was advanced sufficiently to provide a suitable alloy for type metal and techniques for cutting matrices, constructing moulds and casting; the experiments of painters—such as those recorded in the eleventh century by Theophilus Presbyter—had produced oily inks that could be applied to a metal surface and transferred to paper under pressure; and the screw press was already well known. All the prerequisites of typographical printing were thus available. In the decade 1440–50, these elements were combined to make a practical process. The origin of the invention has been much in dispute, but credit must probably go to Johann Gutenberg of Mainz, aided by Johann Fust and Peter Schoeffer.

From Mainz, printing spread to Bamberg and Strasbourg by 1459. During the 1460s it spread further—northwest to Cologne and Marienthal, southwards to Basel, to Augsburg and to Italy. Two Mainz printers, Sweynheym and Pannartz, in 1467 brought the craft to Rome. Also from Mainz, John Spira began work in Venice in 1469, and this town soon became the center of European printing. During the 1470s the craft was taken to France (Paris and Lyons), to Spain (Valencia and Seville), to Budapest, to Cracow, and to England. The biggest printer of the fifteenth and sixteenth centuries, Anton Koberger, set up shop in 1470 at Nuremberg, the commercial hub of central Europe. At the height of his activities he ran twenty-four presses, served by over one hundred compositors, proofreaders, pressmen, illuminators, and binders.

University towns as such had little attraction for printers: learning was no substitute for cash. It was the thriving centers of trading, banking and shipping, the seats of secular and ecclesiastical courts, to which printing spread most rapidly. The newly invented printing press did not immediately abolish the manuscript trade. Many copyists continued their work, and some combined copying with printing: one Arnaldus, for example, printed at least fourteen books in Naples in the 1470s, while maintaining his activity as a copyist. But the immensely greater speed with which printing could produce duplicates inevitably led to its triumph. The cost of a printed book fell to one-fifth the cost of its manuscript.

Venice in the fifteenth century was the outstanding European center of commerce and industry, so it is no surprise that it was in this town that printing first developed in a businesslike way as an aid to scholarship. In the decade 1470–80, we know of no fewer than fifty typographers (almost all Germans) who worked there. But the first of a long line of "master printers" of Venice was a Frenchman, Nicolas Jenson. He became a master of the French royal mint, and in 1458 was sent to Mainz to learn the new art of cutting punches and letters. This opportunity seems to have caused Jenson to take up printing professionally. At any rate, we next hear of him in Venice. He became famous as a type designer—developing the Roman type which we now use in preference to the Gothic style favored by the early German printers. In addition, he was a careful publisher, employing competent scholars to edit his texts. The 150 works he printed cover nearly every aspect of the intellectual activities of his day. In 1471 Jenson published some Latin medical works, and in the following year *Scriptoria rei rustica*, a book that included the Roman agricultural writings of Cato, Varro, Columella and Palladius.

In 1476 Erhard Ratdolt, a German, printed at Venice an Italian version of the *Calendarium* of Regiomontanus, a contemporary astronomer and mathematician. But the main achievement of Ratdolt was the publication in 1482 of Euclid's *Elementa geometriae*, the first Latin edition. Even with more modern facilities, the type composition of mathematical works is a specialized job. For the early printers, books in this field were severe tests of ingenuity. Ratdolt was the first to solve the problem of producing mathematical diagrams typographically. He was also probably the first printer to employ several colored inks simultaneously, and certainly the first to produce colored astronomical diagrams.

The third great printer of Venice was Aldus Manutius. Himself a scholar, Aldus was interested in printing as a means of providing books of use to other scholars. As a student, he became enthusiastic about the masterpieces of Greek literature that were being brought in manuscript from Constantinople to the libraries of Italy, and in his fortieth year, 1490, he went to Venice to establish a printing press. Four years of careful preparation passed before the first product of the Aldine press appeared. The typical Aldine was not the ponderous folio to which earlier printers had been addicted, but a handy pocket volume. His first important publication was the first Greek edition of Aristotle's works, begun in 1495 and completed three years later. The Greek type used by Aldus was not as clear and graceful as Jenson had designed, but he did introduce another innovation, the italic letter.

Printing was brought to England by William Caxton, a Kentish wool merchant. His work led him to Bruges and Cologne, and in one of these cities in 1471–72 he learned the printing art—primarily, he tells us, in order to distribute to his friends an English translation he had recently made of the *Receuil des histoires de Troies*. Some time between 1474 and 1476, at

Bruges, Caxton issued his translation, the first book to be printed in English. Soon after, he set up a printing press near Westminster Abbey, and by 1477 had produced the first dated English printed book, *The Dictes or Sayengis of the Philosophers*. This work has an interesting link with an earlier personality in the history of communication. It had been translated in 1473 from a French work, which had itself been translated from the Latin some years before. The Latin version had first been made in the thirteenth century from an Arabic original, and the latter, though itself dating from the eleventh century, was based on a compilation of Greek philosophy made two hundred years before by the head of the House of Wisdom himself—Hunain ibn Ishaq. Thus Baghdad passed on the classical heritage to London.

Caxton brought out the first editions of the great medieval English authors—Chaucer, Gower, Lydgate, Malory—and seventy-four of the ninety-four books he printed were in English. One of his prefaces gives a picture of the situation faced by these early printers.

> When I resolved upon setting up a press in Westminster, I knew full well that it was an enterprise full of danger. For I had seen my friend Colard, printer of Bruges, fly from the city in poverty and debt, and I had seen Melchior of Augsburg dying a bankrupt, and heard how Sweynheim and Pannartz in Rome had petitioned the Pope for help. Yet I thought that the quick production and the cheapness of books would cause many to buy them who hitherto had been content to live without the solace of poetry and romance. Again, I thought that there are schools and colleges where books must be studied, and I hoped that they would find it better to print than to copy. And there are religious houses, where they are for ever engaged in copying . . . surely, I thought, it will be better for the good monks to print than to copy. But I forgot that there was a great stock in hand of manuscript books, in every monastery and in every college, and these must first be used up before there would be question of replacing them with printed books. . . . And I found, moreover, to my surprise, that there were many great lords who cared nothing for cheapness, and who scoffed at my woodcuts compared with the illuminations in red and blue and gold which adorned their manuscript works. He who would embark upon a new trade must reckon with those who make their livelihood in the old trade. Wherefore my art of printing had many enemies at the outset and few friends. So that the demand for my books has not been found equal to the number which I have put forth. . . .

The central figure of the early book trade was the printer, who chose the manuscripts he wished to print and edited them, carried out the printing, sold the copies, and kept accounts. But the idea of forming a publishing company soon occurred. For example, in Milan in 1472 a priest, a schoolmaster, a professor, a lawyer, a doctor and a printer formed such a company. The first man to make publishing (and bookselling) his exclusive occupation is said to be Johann Rynmann of Augsburg.

FIFTEENTH-CENTURY BOOKS

What effects did the invention of printing have on scholarly communication? First, it greatly multiplied and cheapened the number of copies of each work that were available to scholars and to other interested readers, and so created the possibility of a much wider diffusion of knowledge. Second, the copies so produced were much more uniform than the old manuscript copies, so that standard texts were available. The variant manuscripts of a work could now be collated by one editor, and his definitive text, when printed, was no longer subject to the mistakes of reproduction. Of course, definitive editions of classical works were not achieved at once by early printers—for these we have in general to turn to the results of nineteenth- and twentieth-century scholarship—but printing offered from the start the possibility of standardization.

A further service which printing offered was that multiple identical copies of an original drawing could now be disseminated as woodcuts or metalcuts, and illustrations became an important part of the book. The first printed books to contain illustrations appeared in about 1460, produced by Albrecht Pfister at Bamberg. The simple line drawings were colored by hand. Conrad of Megenburg's *Buch der Natur* was printed at Augsburg by Johann Bämler in 1475, with twelve folio plates of animal and plant pictures. Some ten years later, Peter Schoeffer of Mainz published a German herbal with nearly four hundred pictures. The first woodcut illustrations printed in several colors were the work of Erhard Ratdolt, as already mentioned. Intaglio engravings on copper plates were also employed. The fine detail that could be achieved by the engraver made these plates especially suitable for maps and charts, and in fact one of the first books to be thus illustrated was the *Cosmographia* of Ptolemy. This was printed by Lapis at Bologna in 1477, containing twenty-six maps engraved by Taddeo Crivelli.

Of the thirty-five thousand editions printed in the fifteenth century, over three quarters were in Latin. Religious texts predominated—45 percent of the whole. There were many editions of Latin grammars. Both the *Consolations of Philosophy* of Boethius and Aesop's *Fables* were very popular. It was not until 1469 that a scientific book of any importance came from the printing press—Pliny's encyclopedia being then printed at Venice and Rome. In the 1470s, a number of Latin medical texts first appeared—books by Arab authors such as Ibn Sina and Hunain ibn Ishaq and by medieval physicians. Hippocrates and Galen first appeared in the *Articella* (Padua 1476), and texts by Celsus and Dioscorides in 1478 (all these in Latin). The same decade saw first editions of Mela's *Cosmographia* (Milan 1471), the agricultural book of Pietro dei Crescenzi (in Latin, Augsburg 1471, and in Italian, Florence 1478), and a collection of Latin agricultural texts (Venice 1472). The first work of Aristotle to be printed was *De animalibus* (in Latin, Venice 1476). In the 1480s, medical first editions in Latin continued to appear, and a number of herbals. Euclid's *Elementa geometriae* was printed in Latin by Ratdolt (Venice 1482) and the botanical works of Theophrastus also appeared (in Latin, Treviso 1483). The commercial arithmetics of Chiarini and Borghi were first printed at Florence in 1481 and Venice in 1484. The closing decade of the fifteenth century saw the publication of the Aldine Aristotle.

Among these thirty-five thousand incunabula editions, Klebs has listed about three thousand as "scientific"—i.e., as coming within the fifteenth-century conception of science. These books were the product of over 650 authors—more than half of whom, as might be expected, were contemporary with the press. Of the remainder, about 100 were classical authors, 25 Arab and 150 medieval (thirteenth to fifteenth centuries). The contemporary authors were largely Italian or German—living near the main centers of printing. In an analysis of Klebs' list of scientific incunabula, Sarton showed that by far the most popular writer was the medieval encyclopedist Albertus Magnus, credited with over 150 editions, closely followed by the works ascribed to Aristotle (both genuine and spurious) and his commentators. Among medical writers, Hippocrates and al-Razi were the most popular, and the Salernitan compilation *Regimen sanitatis* is known in seventy-five editions.

The first contemporary scientist to use the new art deliberately for scientific purposes was the German mathematician and astronomer Regiomontanus (Johann Müller of Königsberg). The need to correct errors discovered in astronomical tables led him to a study of Ptolemy's *Almagest*, and to the writing of the first Western treatise on trigonometry. In 1471, Regiomontanus settled at Nuremberg, and there he and his patron, Bernhard Walther, built the first European observatory and set up a printing press. Until his early death in 1476, Regiomontanus issued astronomical calendars, the regular publication of which was then taken over by Ratdolt.

Finally, a note on printing in the vernacular. The archbishop of Mainz in 1486 issued the following mandate:

> Notwithstanding the facility given to the acquisition of science by the divine art of printing, it has been found that some abuse this invention, and convert that which was designed for the instruction of mankind to their injury. For books on religion are translated from Latin into German, and circulated among the people. Can it be asserted, that our German language is capable of expressing what great authors have written in Greek and Latin? Certainly it cannot. But since this art was first discovered in the city of Mainz, and we may truly say by divine aid, and is to be maintained by us in all its honour, we strictly forbid all persons to translate, or circulate when translated, any books upon any subject whatever from the Greek, Latin or any other tongue into German, until, before printing, and again before their sale, such translations shall be approved by four doctors herein named, under penalty of excommunication, and of forfeiture of the books, and a fine of one hundred golden florins.

All such attempts to hold back the use of national languages were soon to be swept away. Already in Germany, in 1475 Conrad of Megenburg's *Buch der Natur* had been printed at Augsburg, and Marco Polo's travels appeared in German translation at Nuremberg in 1477. In the year before the archbishop delivered his edict, Peter Schoeffer had at Mainz printed a German herbal. And in fact by the end of the century nearly a quarter of the incunabula had been printed in vernacular languages.

The invention of printing, because it resulted in a rapid increase in both the supply of and demand for books, led to a much greater need for lists of them for various purposes, and so a new field of scholarship—bibliography—came into existence. The earliest lists of printed books came, naturally enough, from the men who printed and sold them. Peter Schoeffer of Mainz and Johann Mentel of Strasbourg each issued in 1469 a list of books for sale, and in the following year Sweynheim and Pannartz listed their publications in a *Registrum librorum impressorum Romae*. Apart from such trade lists, scholarly bibliography began with the publication at Basel in 1494 of the *Liber de scriptoribus ecclesiasticus* by Johann Tritheim. This surveyed a thousand authors in chronological order—seven thousand works in all—with an alphabetical index of authors (the index still followed the medieval practice of listing authors by their forenames). Tritheim, wrote Besterman, brought to bibliography "a true love and knowledge of books, the practical experience of a librarian and cataloguer (at his abbey in Spannheim), and something of that inexplicable ardour for system, for order, which is so important an ingredient in the make-up of the bibliographer."

Section 4

The Scientific Revolution (1450–1700)

THE RENAISSANCE

In the sixteenth century, science began to move rapidly forward. Greek and Arab knowledge had been absorbed, and was now to be surpassed. A characteristic feature of the Renaissance was its boundless appetite for new knowledge, new experience. Its leading spirits were interested in everything. As Leonardo da Vinci wrote to the Duke of Milan:

> In the art of war, I have a process for the construction of very light bridges. . . . I know how to drain moats. . . . I have a means of destruction by mining. . . . I know also how to make light cannon. . . . I know how to construct secure and covered wagons. . . . In short, I am able to devise endless means of attack. . . . In time of peace, I believe that I can compete with anyone in architecture, and in the construction of both public and private monuments and in the building of canals. I am able to execute statues in marble, bronze and clay. In painting, I can do as well as anyone else.

Salomon's House of Wisdom, in Francis Bacon's *New Atlantis*, sought "knowledge of causes and secret motions of things, and the enlarging of the bounds of human empire, to the effecting of all things possible."

The Renaissance wanted to experience "all things possible"—to taste the whole of life, to feast on it, even to satiety. "Then did they fall upon the chat of victuals, and some belly furniture to be snatched at in the very same place. Forthwith began flagons to go, gammons to trot, goblets to fly, great bowls to ting, glasses to ring. . . . I sup, I wet, I humect, I moisten my gullet,

I drink, and all for fear of dying. Drink always and you shall never die." This is the spirit of the opening scene of *Gargantua and Pantagruel*, the spirit which surges through the whole of Rabelais, the spirit of the Renaissance: the flaming vitality, the intoxication of recaptured learning, the Columbus-like voyages of the mind. If the weather was rainy, young Gargantua

> studied the art of painting or carving . . . examined passages of ancient authors . . . went to see the drawing of metals, or the casting of great ordnance: how the lapidaries did work, as also the goldsmiths and cutters of precious stones. Nor did he omit to visit the alchymists, money-coiners, upholsterers, weavers, velvet-workers, watch-makers, looking-glass-framers, printers, organists, and other such kinds of artificers, and visited the shops of druggists, herbalists and apothecaries, and diligently considered the fruits, roots, leaves, gums, seeds, the greases and ointments. . . .

Encyclopedic thirst for knowledge was not a phenomenon peculiar to the Renaissance. Many examples have been recorded in our story—Aristotle, Pliny, and a galaxy of medieval encyclopedists. But all these men, even Aristotle, were mainly engaged in recording existing knowledge. They could prepare, over the years, a vast compilation of all that was known, fitting innumerable facts into a systematic scheme. Their work was conceived as a whole, and could be published as a whole. The Renaissance scientist, however, burst out of all such scholastic systems. His mind ran in every direction, ferreting out new facts of nature, experimenting, devising new theories, throwing up a chaotic mass of new knowledge which did not fit into any existing system. Leonardo covered thousands of pages with notes of his new observations and ideas. Francis Bacon listed hundreds of projected new "histories."

The pace of industrial development had begun to quicken. Mining and metallurgy had a long history. Technical advances now occurred, and the scale of manufacture increased: for example, two of the largest English industrial organizations of the time were the Society of the Mines Royal and the Mineral and Battery Works. Large enterprises were also formed in shipbuilding. Sugar refining, the manufacture of salt, of saltpeter, and of alum, and glassmaking were other developing industries where machinery in small factories required considerable capital outlay. Entrepreneurs often received "patents of monopoly" from the Crown. Industry was developing also in other countries, for example Italy. Perhaps the largest single enterprise in Venice was the state shipyard building naval and commercial vessels. Alum—needed for glass manufacture, leather tanning, and as a mordant in dyeing—was also an important product.

Not the least remarkable feature of the writings of the advocates of the new experimental science was their unanimous conviction that even vaster new empires of knowledge were waiting to be conquered, and that the conquest required the mobilization of as many heads and pairs of hands as could possibly be brought together. Nature was seen to present myriad problems, each requiring a solution, so many that concerted action was imperative. Every individual scientist felt the need to share his latest discoveries with his fellows, to benefit by their criticisms and suggestions, to learn as quickly as possible of the advances of others, and to enjoy the stimulus of constant intercourse with like minds. A loose-knit European-wide network of natural philosophers was built up in the sixteenth century.

In general, men of science came from well-off families of merchants, lawyers and others. Some were still churchmen, but others were doctors or professors. Some professors set up feepaying courses of instruction in their own houses, independent of universities. Other scientists depended on the patronage of sovereigns, dukes or cardinals, as did the newly emerging engineers such as Leonardo, who worked for Italian dukes and King Louis XII of France. The rapidly growing trend to the vernacular languages, fostered by the growth of printing, made the new learning accessible to people outside the academic world. There was an increasing tendency for those with practical skills (engineers, pharmacists, metal workers, surveyors, navigators, etc.) to write about their work. Some outstanding technical books were produced, e.g. the *Pirotechnia* of Vanocchio Birungiccio (1540), *De re metallica* by Georgius Agricola (1556), and the *Beschreibung allerfürnemisten mineralischen Ertzt* of Lazarus Ercker (1574).

THE INTELLIGENCERS

Many Renaissance scientists satisfied their thirst for news by voluminous correspondence. "It was one of the important achievements of the Renaissance," writes Raven, "that it restored to the learned world the habit of letter-writing in the mode of Cicero, letter-writing upon subjects of technical and professional interest." At first Latin, the universal language of the learned, was used, and it continued as one medium for scientific communication into the eighteenth century, even though the vernaculars steadily grew in importance.

The problem of delivery, though difficult, could be overcome by the agency of merchants. Big financial houses, like the Fuggers based in Germany, and leading printers, sent regular mail-bags from market to market, and private correspondence was readily accepted by them for transmission. As an instance of their service we read that on August 9, 1516 one Ulrich von Hutten wrote from Bologna to Richard Croke at Leipzig University, asking him for a book: on August 22 he wrote acknowledging its receipt. It was not until the seventeenth century that regular postal services were organized. In 1633 a weekly mail was set up between London, Antwerp and Brussels. Soon regular services were running, for example, on all the main roads of England.

Some of the merchants became interested in science themselves and through their foreign contacts built up a circle of scientific correspondents. The "intelligencer" acted as a liaison between the scientists of his own country and those abroad. But merchants were not the only ones to perform this function. An early example was the sixteenth-century English scholar John Dee. As a young man, he traveled in Holland and France. He corresponded with men in the universities of Orleans, Cologne, Heidelberg, Strasbourg, Verona, Padua, Ferrara, Bologna, Urbino and Rome. Later in life he traveled to Antwerp, Bremen, Lubeck, Cracow and Prague.

One of the most industrious correspondents of the sixteenth century was the biologist Conrad Gesner of Zürich. In the preface to *Historia animalium*, for example, he listed fifty-two scholars who had helped him by letters or discussion. By correspondence, Gesner collected together the isolated observations of many collaborators, which he combined with his own observations. For example, he recorded descriptions of birds supplied by his friends Tschudi in Switzerland, Jacques d'Alechamps at Lyons, Guillaume Rondelet at Montpellier, Lammergeier in Austria, and a friend at Strasbourg. He corresponded with the physician John Caius at Cambridge, with friends at Danzig, Locarno, Leipzig and Cracow. As Raven points out, such correspondence made available "masses of knowledge which until then would have been jealously guarded, and encouraged a full exchange both of experience and of material. . . . In his sense of the unity of knowledge, of the need for cooperative effort, Conrad Gesner anticipated and by his heroic labours facilitated later developments."

Moving onward in time, let us look at Paris in the 1630s. Rich lawyers and state counselors were setting up cabinet displays of natural history, and also libraries, and assembling their friends to discuss scientific matters. Notable were the "conferences" of Marin Mersenne. These were regular discussion meetings of scholars both French and foreign. Friar Mersenne, friend and follower of Descartes, was during this period the so-called "secretary-general of European scholarship." He corresponded with everyone of note in science—Huygens the Dutch physicist, Galileo and Torricelli the Italians, Hevelius the Danish astronomer, Hobbes the English philosopher, the French chemist Jean Rey, the French mathematician Fermat, and many more.

Mersenne firmly grasped that the advancement of science required collective work and that, though specialization was necessary, science formed an inviolable unity, that scholars should seek to acquire a general understanding of the whole field. He did more than correspond with the great—he encouraged the humble. "There are often," he wrote, "men in small towns who surpass us all in knowledge, particularly in sciences where they have relied on their own discoveries rather than on books," and he diligently sought out such men. He was interested not only in "philosophical" scientists but also in craftsmen: "If artisans learn to love science, they will be able to help the philosophers in many ways, particularly in providing them with an understanding of their normal activities."

One of Mersenne's most active correspondents was Theodore Haak, a German who came to England in 1625. Haak was the friend of several English scientists, such as Gabriel Plattes the agricultural writer and John Pell the mathematician, and acted as a "corresponding secretary" for his group. Exchanges of books occurred: for example, Haak sent over to France a manuscript of Pell's *Idea of Mathematicks*, in which he suggested the publication of a compendium summarizing all extant mathematical books. Mersenne sent back criticisms by Descartes and himself which were included in Pell's printed work. Mersenne maintained that it would be better to make a selection of a dozen of the best works in each branch of mathematics, presenting all that was good, without caring about the rest.

Another member of Haak's group, Samuel Hartlib, later became known as "the great intelligencer." He was born in Prussia, and came to London in about 1628. Repeatedly in letters of the period his help is acknowledged—he was clearly a most useful man, "a man very well known, beloved and trusted by all sides, a man exceeding painful and diligent," as a contemporary wrote.

Another influential intelligencer was Henry Oldenburg, a German who came to England in the 1640s and again in 1653. The second journey was to negotiate on behalf of the city of Bremen with Oliver Cromwell, a mission which introduced him to John Milton, then Latin secretary for the government. At that time, Milton wrote to him, "You have indeed learnt to speak our language more accurately and fluently than any other foreigner I have ever known." In 1654 Oldenburg settled in England permanently, as a tutor to the nephew of the chemist Robert Boyle. Thus began a friendship which later proved most valuable to Boyle, who rarely lived in London, for Oldenburg reported regularly to him the scientific news of the capital. Oldenburg also maintained a large correspondence with friends abroad, passing on European news to his patron, and during 1664–67 he saw to the printing in English of seven of Boyle's books, and translated five of them into Latin.

Other countries joined in the interplay of scientific activity. Queen Christina of Sweden, from her accession to the throne in 1644, began to invite foreign scholars to her court. Again, Holland at this time was a refuge for many scholars displaced from their own countries for religious or philosophical reasons. They congregated at the universities, particularly the University of Leyden, founded in 1575. They maintained many contacts with their colleagues in other countries, who often visited them. The "scientific tour of Europe" was one of the best ways of learning the latest ideas. At Leyden, the publishing house of the Elzévirs was active (founded 1583), and scholars throughout Europe sent their manuscripts to them for printing.

THE FIRST SCIENTIFIC ACADEMIES

That Italy should be the home of the first scientific academies is hardly a matter for surprise. It was one of the most highly urbanized regions. The Italian cities such as Milan, Venice, Naples and Palermo were among the largest and richest in Europe. They were economically advanced, and had higher levels of literacy and perhaps numeracy, with a consequent heightening of interest in science. Moreover, academies outside the university were already known in Italy. The humanists had found that there was no place for their literary interests in the universities, which were still staunchly defending the medieval Aristotelian traditions, so loosely grouped humanist academies had been set up, where the Greek language was taught and where newly discovered Greek manuscripts could be studied, often with the assistance of refugees from Byzantium. We have mentioned Marsilio Ficino's so-called Platonic academy at Florence as perhaps the most influential.

These humanist groupings were essentially literary. If natural philosophers were to meet, they would have to create separate organizations. The first known of these was the Accademia Secretorum Naturae, founded in Naples in 1560 by the alchemist Giovanni Baptista della Porta. Perhaps it is stretching the point to call this a scientific academy, for its members seem mainly to have been dabblers in what was termed "natural magic," an interest in sensational and mystifying experiments. The society came under suspicion of witchcraft and heresy, leading to its dissolution.

At the beginning of the next century, the Accademia dei Lincei was founded in Rome by the duke Federigo Cesi, himself a keen experimenter, collector of natural curiosities, beekeeper and botanist. The first members were men with very similar interests. Their regular meetings and correspondence were sufficient to bring them under suspicion, and for a while the group was prudently disbanded. However, in 1609 it was formed anew, this time a larger body that included both della Porta and Galileo. It aimed to establish a scientific brotherhood in "non-monastic monasteries" in all parts of the world, each with museum, library, printing press, laboratory and botanic garden. Though it did not succeed in this, the academy played a useful part in the advancement of experimental science by acting as a center for research. Its experiments were recorded in the *Gesta Lynceorum*, which may be regarded as the first publication of a learned society.

In 1657, Duke Ferdinand of Tuscany organized the Accademia del Cimento, bringing together formally a group of scientists and engineers who had for some years been in receipt of his patronage. Although it endured only ten years, it was widely influential, being the first society to undertake collective work, and an account of its work was published in 1667.

In England, the inability of the universities to provide the education necessary to professions other than the church provoked the foundation of rival institutions even in the fourteenth century—for example, the collegiate Inns of Court for lawyers. In 1368 the master-surgeons of London formed a guild and later, together with the physicians, petitioned for royal recognition. Eventually, in 1518 the Royal College of Physicians was incorporated (a few years earlier, a Royal College of Surgeons had been formed in Edinburgh). But most English scientists during the sixteenth century worked outside any formal organization. Some put forward schemes for new colleges and institutions to teach their subjects—for example, Humphrey Gilbert, stepbrother to Walter Raleigh, in 1570 presented proposals to Queen Elizabeth for an academy to meet the requirements of the new age, but nothing came of them.

Thomas Gresham, a wealthy city merchant, not only had a plan, but also the money to back it. His will endowed a college in the city of London, and as a consequence Gresham College was opened in 1598. Salaries were provided for professors of law, physic, rhetoric, mathematics, astronomy, geometry and music. Henry Briggs, inventor of logarithms, was the first professor of mathematics. The Gresham professors—at the center of mercantile London—eagerly enlarged their horizons by contact with merchants, sailors and instrument makers. When the Royal Society was later formed, it was only natural that the college should be its first home.

The mention of instrument makers is evidence of new links between science and craft knowledge. Precision mechanisms had already developed in the fourteenth century in the form of clocks, and at the same time the making of astronomical instruments—neglected since the days of Ptolemy—revived at Oxford and Paris. In the next century Nuremberg and Augsburg became centers of instrument making, and early in the sixteenth century the craft spread to England. By 1650 there were at least thirty instrument makers in London, many of whom had close connections with science—indeed, by 1776 there were half a dozen instances of instrument makers becoming members of the Royal Society.

The theoretician in chief of the progressive movement was Francis Bacon, Lord Verulam, Viscount of St. Albans, whose series of books was intended to form a single integrated work to be known as "the great instauration," a comprehensive plan for the reform of society. The *Advancement of Learning* (1605) and *Novum organon* (1620) played their part in trenchant criticism of the old university learning and in advocating an experimental science developing in close contact with industry. His unfinished *New Atlantis* of 1627 gave a description of a utopian scientific academy.

Bensalem is an island utopia in which travelers are shown Salomon's House, which might be termed a research institute. In addition to parks, animal enclosures, gardens, orchards and fish pools, there are buildings for all kinds of operations: a furnace house, a perspective house for optical experiments, sound houses, perfume houses, houses with "engines and instruments for all sorts of motions," a mathematical house, and "houses of deceits of the senses." An account of the "several employments and offices of our fellows" is then given. Twelve sail into foreign countries, "who bring us the books and abstracts and patterns of experiments . . . these we call merchants of light." Three "depredators" cull experiments from the books; three "pioneers" carry out new experiments. The results of all this activity are collected into tables by three "compilers." Three more fellows ("benefactors") attend to the practical application of the experimental results. After due consideration, fellows known as "lamps" design new experiments, more penetrating than those already carried out, and lastly, "we have three fellows who raise the former discoveries by experiment into greater observations, axioms and aphorisms. These we call interpreters of nature."

In his own words, Bacon "rang the bell to call the Wits together." He widely influenced his contemporaries and the following generation, in which discussions began that eventually led to the formation of the Royal Society of London.

THE ROYAL SOCIETY

The intelligencer, Samuel Hartlib, had much personal interest in educational reform, and was in close contact with like-minded people concerned with the betterment of society, such as John Dury (author of *The Reformed Librarie-Keeper*). This group invited the Czech educationist Comenius (Jan Komensky) to England in 1641 to propagate his views on university education, but during the struggle between king and parliament then proceeding, few people had thought to spare for educational reform. Hartlib's circle continued to develop its ideas and in 1645 began to discuss and publicize a proposal for an "Office of Address," which would serve as an international correspondency for the advancement of learning, and direct the efforts of inventors.

Within a few years, the proposal had been refined into an Office of Address for Communications, which would be "a centre and meeting place of advices, of proposals, of treaties and of all manner of intellectual rarities." An Agent for Communications would encourage English scholars and keep up an international correspondence. It was particularly important for him to employ the librarians of the major libraries to supply useful information from their collections of books and manuscripts. By 1655, Hartlib's "agency for the advancement of universal learning" was being conceived as a national institution headed by an agent and a council of trustees, with two permanent secretaries employed on foreign and internal correspondence, employing inventors and experimenters, building up a museum and establishing a laboratory, printing and publishing important books and useful information. A number of the founders of the Royal Society (such as Robert Boyle the chemist, William Petty the economist, John Evelyn the diarist, John Wilkins and Henry Oldenburg) were associated with the plans for the agency. At about this time, the College of Physicians was giving an example of collaborative scientific work: in 1648, a chemical laboratory had been set up, and cooperative medical researches were being undertaken.

In 1645 a group led by John Wilkins began to hold regular meetings in London (often in the rooms of the professor of astronomy at Gresham College), where, as the mathematician John Wallis wrote, "we confined ourselves to philosophical inquiries, and such as related thereunto: as physick, anatomy, geometry, astronomy, navigation, staticks, mechanicks, and natural experiments. We there discoursed the circulation of the blood, the valves of the veins, the Copernican hypothesis, the nature of the comets and new stars, the attendants on Jupiter, the oval shape of Saturn, the inequalities of the moon, the several phases of Venus and Mercury, the improvement of telescopes. . . .” Later, some members of the group, such as Wilkins and Wallis, moved to Oxford, where they were joined by the architect Christopher Wren and Robert Boyle. Groups continued to meet in both London and Oxford.

In 1660, after a lecture in London by Christopher Wren (then Gresham professor of astronomy), the group discussed the formation of a “college for the promotion of physico-mathematical experimental learning.” They were soon after informed that King Charles II had approved of a scheme to form a society, and they received a charter for the Royal Society for the Improvement of Natural Knowledge in 1662. The name Royal Society appears to have first been used by John Evelyn, in the dedication of his translation of Gabriel Naudé’s *Avis pour dresser une bibliothèque*, 1661. The first appointed secretary was Henry Oldenburg, and Robert Hooke (an assistant to Robert Boyle) was appointed as the “curator of experiments.” Thus scientists in England at last achieved official recognition.

EUROPEAN ACADEMIES

In 1600 the university of Paris had thirty thousand students and great influence, but the medieval curriculum was still in place. Teaching was only in Latin, censorship of books was very strict, modern views such as those of Descartes were excluded. Outside the university, the Collège de France had been founded in 1518 to teach humanism, and subsequently chairs of medicine, mathematics, surgery and botany were introduced. Of the provincial universities, only the medical center Montpellier still retained its reputation.

As in Italy and England, scientists began to meet informally—we have noted meetings at the residence of Mersenne, but they took place also at the homes of many rich patrons. In the 1640s, Pierre Bourdelot, physician to the prince de Condé, started regular meetings. In the 1650s, meetings began at the house of Habert de Montmort, but never achieved harmony of purpose. Its secretary, Samuel Sorbière, pointed out some of its difficulties: “to build an arsenal of machines to carry out all sorts of experiments is impossible . . . only kings or a few wise and rich republics can undertake to erect a physical academy where there would be constant experimentation.”

Colbert, the minister in charge of finance, industry and the navy, was quick to notice that scientists were paying attention to industrial needs and problems, and in 1662 gave support to a scientific society in Caen, led by

Daniel Huet and André Graindorge, who corresponded regularly with Oldenburg. When the Royal Society of London was founded, Colbert saw that French science also needed official recognition. He therefore proposed to Louis XIV that the scientists now meeting informally should be incorporated in an official academy, similar to the Académie Française (which had been founded in 1635 as a literary academy). The Académie Royale des Sciences was formed in 1666. It was allotted rooms in the king's library, where experimental work could be collectively planned, performed and discussed.

In Germany during the seventeenth century, several learned societies were founded, at first mainly literary and philological. Among these was the short-lived Societas Ereunetica in Rostock, founded in 1622 by the biologist Joachim Jung, which included the fostering of science among its aims. In 1652 Johann Bausch, a physician of Schweinfurt, founded the Collegium Naturae Curiosorum to further the publication of scientific papers and monographs. In 1677, the emperor Leopold became patron. It was not until 1700 that Germany had a scientific society that could compare with those in England and France, when the Societas Regia Scientarum was constituted in Berlin by King Frederick I. Plans for such a body had long been put forward in memoranda by Gottfried Leibniz, philosopher, mathematician and librarian.

The formation of academies had the consequence that a new source of support became open to scientists: no longer were they each individually dependent on the wealthy patron. The academies could seek financial support from the state and the wealthy and disburse it to individual scientists. A collective administration of science by scientists began to emerge.

THE DEVELOPMENT OF JOURNALS

As the seventeenth century continued, the number of scholars meeting regularly for discussions in London, Paris, Venice, Rome, the Hague, Berlin and elsewhere increased. The "intelligencers" and "correspondents" who kept in touch with other groups spent much time copying the same scientific news and views for each of their widely scattered contacts. The need was clear: the use of the printing press to provide newsletters.

Outside science, examples of this use were well known. The Venetian republic in the sixteenth century began issuing *coranti*, *gazetti* or *foglietti*, official bulletins with news of current events. The earliest known periodical newspaper of definite continuity was the *Mercurius Gallo-Belgicus*, which started in Cologne in 1594, and was current till about 1630. Soon there were scores, perhaps hundreds, of such "mercuries," each prepared in an "intelligence office."

Scientists of the time saw a learned journal as enabling the publication of short papers about new discoveries, observations and experiments. Remarks, attacks and replies could be noised abroad throughout the learned world, giving opportunities for continued discussion. Corrections, denials, requests for information, and announcements of projects could find a place. "By their criticisms and threats of exposing plagiaries," writes Barnes, "by removing scholars from the condition of privacy and isolation, by the vast panorama of objective learning which journals spread before savants, learned periodicals contributed to the sum total of the critical element in the intellectual life of Europe."

In 1663 the historian Mézeray took out letters patent in France for a *Journal général de littérature*, which was to include scientific and technical news, but his project was not realized. The first editor of a scientific journal was to be the French lawyer Denis de Sallo, a friend of Colbert. He was an omnivorous reader and kept two secretaries to compile notes and make extracts from the books he read. It occurred to him that he might do for the public what he did for himself. Colbert approved of the scheme, and in January 1665 the first issue of the weekly *Journal des sçavans* appeared. "The journal," wrote de Sallo, "has been invented for the relief of those either too indolent or too occupied to read whole books. It is a means of satisfying curiosity and of becoming learned with little trouble." The reader was advised that the journal would supply (1) a catalogue and short description of new books, (2) obituaries of famous men, (3) news of experiments and observations in physics, chemistry, mathematics, astronomy and anatomy, and of useful machines, (4) legal and ecclesiastical judgements, and (5) current events in the world of letters. After thirteen issues, the journal was suppressed on account of Jesuit criticism of its radical views, but resumed publication in 1666 under the safer editorship of the Abbé Gallois. During the first two years, about two-thirds of the journal was devoted to scientific matters.

Now we turn to England. "If all the books in the world except the *Philosophical Transactions* were destroyed," wrote the nineteenth-century scientist Thomas Huxley, "it is safe to say that the foundations of physical science would remain unshaken, and that the vast intellectual progress of the last two centuries would be largely, though incompletely, recorded." This was the tribute paid to the oldest serial publication of a learned society that has continued to the present time.

The *Philosophical Transactions* of the Royal Society of London began under the society's license as the private venture of its secretary Henry Oldenburg. In November 1664, Oldenburg received a letter from Henri Justel of Paris—himself a considerable intelligencer—offering to correspond regularly. This began an exchange of news that continued until Oldenburg died in 1677. In his first letter, Justel mentioned de Sallo's plan to publish a journal, and invited Oldenburg to contribute, but the latter was too involved in Royal Society work to take up the offer. He said of himself: "Attends meetings . . . noteth the observables said and done there . . . digesteth them in private . . . takes care to have them entered in the Register . . . solicits performance of tasks recommended . . . writes all letters abroad . . . disperses far and near store of directions, inquiries for the society's purpose. Query, whether such a person ought to be left unassisted."

In February 1665, Sir Robert Moray, a stalwart of the Royal Society, wrote to a French colleague that a sample of the *Journal des sçavans* had been seen in London, "but already we have found things to criticise in it. . . . Mr Oldenburg has shown us a sample of a similar plan, much more philosophical, and we hope to get him to begin it, if it can be done." It was done, the first number of the *Transactions* appearing in March 1665. The issue included an account of the improvement of optic glasses at Rome; observation of a spot on one of the belts of the planet Jupiter; notes on an experimental history of cold; accounts of a "very odd monstrous calf"; a peculiar lead ore; and a catalogue of books written by the mathematician Pierre de Fermat, lately deceased. Of the ten items in this number, three came from the French journal. In the tenth number, Oldenburg stated that the *Transactions* were intended "not only to be (by parcels) brief records of the emergent works and productions of the universe; of the mysteries of nature; of later discoveries; and of the growth of useful inventions and arts; but also and chiefly, to solicit in all parts mutual aids and collegiate endeavours for the farther advancement thereof." Oldenburg edited the journal till his death, when it continued under the private editorship of later secretaries of the Royal Society. In 1752, it was taken over by the society, to be published by its Committee on Papers.

The French and English journals described above were soon followed by other publications. Michael Ricci, during 1668–79, edited the *Giornale de letterati* at Rome. In 1670, the Leipzig Akademie der Naturwissenschaften began publication of the *Miscellanea curiosa medico-physica*, which under various titles has continued to the present day. The Dutch anatomist Thomas Bartholin founded, for the Societas Medica at Hafnia, the *Acta medica et philosophica* (1671–1829). In Paris, in 1683, Jean Paul de la Roque, at that time editor of the *Journal des sçavans*, started a new periodical for material not suitable for the former, calling it the *Journal de médicine*. In London, the *Weekly Memorials for the Ingenious* was started in 1682, a collection of reviews of current books.

More important than any of these was the monthly *Acta eruditorum*, published at Leipzig, and edited by Otto Mencke, assisted by numerous correspondents in Germany, Holland and England. The articles were written in Latin and dealt with all sides of learning. The first volume (1682) included contributions from the English botanist Nehemiah Grew, Robert Boyle, the English physician Thomas Sydenham, the French physicist Denis Papin, the Italian physiologist Gian Borelli, the Dutch microscopist Anton van Leeuwenhoek, Gottfried Leibniz, the French mathematician Jacques Bernoulli and the German astronomer Johannes Hevelius. The *Acta* became an international forum for the exchange of scientific ideas.

The early editors faced many difficulties. It was a problem to obtain material for publication. Although the journal was developed to accommodate a growing volume of scientific work that needed reporting, many scientists were slow to use the new medium. Since contributors were not plentiful, editors were obliged to print inferior articles—their own or contributed—and many analyses of books. Collaboration in editing became customary by 1700, particularly when the journal was state subsidized, as were the *Acta eruditorum* and the *Journal des sçavans*. In 1693, Cornand de la Crose wrote in the *Memorials for the Ingenious* that "experience has taught me that the perfect establishing of a journal is not the performance of private men."

Let us conclude this section with notes on some sixteenth- and seventeenth-century bibliographers. First must come Conrad Gesner, whose *Historia animalium* has already been mentioned. This Zürich scholar lived less than fifty years, but crowded into his life an enormous amount of scientific and literary labor, acquiring that many-sidedness in learning which was so typical of the great Renaissance figures. Gesner's *Bibliotheca universalis* (1545) alphabetically listed fifteen thousand Latin, Greek, and Hebrew books; his *Pandectae sive partitionum universalium* classified these items under twenty-one heads, subdivided, followed by an alphabetical subject index of twenty-four thousand headings. It has been estimated that the works cited represent 20–25 percent of the then existing books in the languages mentioned. He also prepared zoological, surgical and botanical bibliographies.

These last were not the earliest subject lists of less than universal scope: Symphorien Champier in 1506 published *De medicine claris scriptoribus*, dealing with ancient medical writers and Italian, French, Spanish, German and English authors. Another early medical bibliography was that of the German herbalist Otto Brunfels, the *Catalogus illustrium medicorum* (1530).

Although a quarter of a million book titles had been published by the beginning of the seventeenth century—and a million more by the end—attempts at universal bibliography continued. For example, the *Bibliotheca realis universalis* of Martinus Lipenius was issued in parts from 1679 to 1685, covering 125,000 items.

The most eminent bibliographer of the century was perhaps Cornelius à Beughem of Emmerich. His *Bibliotheca medica et physica* (1681) listed the works of about two thousand writers that had been published during the previous thirty years. In 1685 he prepared a similar *Bibliotheca mathematica et artificiosa*. These works indicated a new need—for lists of currently important scientific books rather than for universal retrospective lists. With the coming of the scholarly journals, Cornelius recognized the need also to record their contents. In 1683 he produced *La France sçavante*, which was in effect a bibliographical index to the *Journal des sçavans*, founded in 1665. His *Apparatus ad historiam litterarium novissimam*, which appeared in installments from 1689 to 1710, supplemented this and enlarged it to include the contents of other journals. His *Syllabus recens in re medica, physica et chymica* (1696) was a similar review of scientific articles in the transactions of contemporary learned societies.

THE PROGRESS OF LIBRARIES

The century following the invention of printing witnessed another development that greatly affected the fortunes of libraries. All over northern Europe, the dissolution of monasteries during the Reformation led to the dispersal or destruction of manuscript collections. Things seem to have gone worst in England. During the years following 1535 about eight hundred monasteries and convents in that country were secularized, and their book collections received the most careless treatment. In 1550 the Commissioners of Edward VI came to Oxford and so completely emptied the university library that shortly after its furniture was sold as useless.

By 1560, private collectors began to work more openly to track down and rescue dispersed manuscripts and collect them into private libraries. Many of these collections eventually found their way into the college libraries of Oxford and Cambridge. A little later, as we have noted earlier,

Thomas Bodley had the idea of reestablishing a library for the University of Oxford, and this library was reopened in 1602. In 1610 Bodley negotiated an agreement with the Stationers Company (the licensed publishers of printed books) entitling the library to receive a copy of every book published in England.

In 1651, the group of scientists that were meeting regularly in Oxford agreed to examine works of scientific interest in the Bodleian library, and the astronomer Seth Ward later commented on the progress of the scheme:

> We have (every one taking a portion) gone over all or most of the heads of natural philosophy and mixt mathematics, collecting only an history of the phenomena out of such authors as we had occasion and opportunity, our first business is to gather together such things as are already discovered and to make a book with a general index of them, then to have a collection of those which are still inquirenda. . . . But because (not knowing what others have done before us) we may probably spend time upon all that which is already done, we have considered it requisite to examine all the books of our library (everyone taking his part) and to make a catalogue or index of the matters and that very particularly in philosophy, physic, mathematics and indeed in all other faculties, that so great numbers of books may be serviceable and a man may at once see where he may find whatever is there concerning the argument he is upon, and this is our present business which we hope to dispatch this Lent.

Whether they ever did "dispatch" it is not known.

Early attempts at founding reference libraries in London were set back by the great fire of 1666. The Royal College of Physicians library had been based on the collection of its founder, Thomas Linacre, and had been augmented by bequests from William Gilbert, William Harvey and other scientists. It included not only medicine but also geometry, geography, cosmography, astronomy, physics and zoology. Except for 140 books, it was destroyed by the fire. Also lost for the most part were the libraries of Sion College (founded 1635) and the Society of Apothecaries (1633).

However, a number of private libraries existed in London during the late seventeenth century: Charles Scarburgh and Jonas Moore each had mathematical collections; Samuel Pepys, at one time secretary to the Admiralty and president of the Royal Society, made an important collection of material on geography and hydrography, compiling a *Bibliotheca nautica* in 1695; Francis Bernard, physician at St. Bartholomew's hospital, had a library of nearly fifteen thousand works; John Moore, bishop of Ely, collected at his house in Holborn some twenty-nine thousand printed books and 1,790 manuscripts, later bought for the university of Cambridge.

In 1708 a book collector, John Bagford, published an account of libraries in and about London. Sion College, rebuilt in 1669, was the only library available to the public within the walls of the city. The Royal College of Physicians had rebuilt their library. Christ's Hospital possessed a "neat" library for the use of masters and students, and St. Bartholomew's Hospital library had been formed in 1677. The Arundel library of the Royal Society was at Gresham College.

Turning to Germany, in a sixteenth-century Meissen chronicle we read: "It is quite common for most of the nobles and burghers, even if they do not actually study, to be able at least to read and write, to bring together in their homes fine libraries of all sorts of good books of godly writings, and to attract to their hearths excellent and profitable historians, physicians and others." One of these burghers, who was called "the richest and most learned man in Germany," built up a library in Munich: this was John Jacob Fugger, a member of the merchant family to whom reference has already been made. Early in the seventeenth century his collection became a copyright deposit library for Bavaria.

Also on the Continent there were dreams of "universal" libraries. Gabriel Naudé, at first a student of medicine, was librarian to various French cardinals. In 1627 he published his *Avis pour dresser une bibliothèque*, which set out a program for a library stocked with all important books (originals and translations), along with commentaries and reference works. In 1642 he entered the service of cardinal Mazarin, and in eight years acquired forty thousand items. Gifts streamed in. Generals and diplomats abroad were given purchasing commissions to carry out. The library was dedicated "to all those who desired to come to study there."

The administration of books, as we have seen, was becoming an important part of scientific communication. There was a need, as Naudé recognized, not only to collect books, but for systematic organization and expert administration of collections running into tens of thousands. The day of the serious research library was dawning. In 1661, Colbert was placed in control of the Bibliothèque du Roi, and took up the tradition of Mazarin. His librarians included the historian Etienne Baluze, who quadrupled the size of the collection in twenty years, and Nicholas Clément, who prepared a classed catalogue in 1675.

During this time of advance, Gottfried Leibniz visited Paris, met Baluze and Clément, and read Naudé. In 1676 he became librarian to the duke of Brunswick, and in later years bombarded local princes with plans for library development, few of which were put into effect. He envisaged a complete, well-administered book collection, comparing it to a gathering of the greatest men of all ages and races who communicate to us their most select thoughts. He clearly presented the importance and characteristic features of the large scholarly reference library.

EUROPE AND THE WORLD

This account of communication in science has so far focussed on the situations in early Mesopotamia, then in lands bordering the Mediterranean, and then proceeding to look at Europe generally. Many general histories speak of events from the fifteenth century on as "the expansion of Europe" or as "Europe discovering the world." These expressions may describe the situation as Europeans saw it, but do not give an adequate picture of interactions between Europe and the rest of the world. The purpose of this section is, first, to glance back at contacts between the Mediterranean, Asia and Africa before the fifteenth century; second, to give a brief account of the geographical explorations that started during the Renaissance and led to European settlements throughout the world; and third, to mention some of the effects on science of the new discoveries. Later sections will consider the development of scientific activities outside Europe.

Much earlier, mention has been made of the roads built by the Persians, running sixteen hundred miles from the Persian Gulf to the Aegean Sea, along which Alexander's armies later traveled. The penetration by Alexander into Mesopotamia, even as far as India, and the founding of cities by his Greek followers established closer relations between these areas and the Mediterranean. By the first century A.D., there was regular trade between Rome and Asia: by land, from Tyre (the old Phoenician port on the Mediterranean) through Persia and on via the "Silk Road" to China; and through the Red Sea and Arabian Sea to India. With the coming of Islam, Arab influence in time spread into North Africa and along the East African coast, as well as into India, a process that continued even into the seventeenth century, thus forming new links between the Mediterranean and countries further east.

Trading enterprise within Europe revived at the end of the eleventh century, the Crusades playing a leading part in this revival. The merchants of Venice were among the first to seize the opportunities offered by the Crusades for extending trade with Constantinople and the eastern Mediterranean. Their trading posts collected Asian spices, silks, carpets and porcelain from Arab traders, and distributed them to western Europe.

Intermittent European contact with Asia thus occurred before the Renaissance—we have only to think of Marco Polo's travels from Venice to China and India in the thirteenth century. He was not the only traveler to Asia during this century: two Franciscan friars (John of Pian and a companion) trekked to Mongolia and back to meet the Great Khan; and they were followed by William of Rubruck. Later, friars such as John of Montecorvino and Odoric of Pordenone reached Peking (at that time ruled by the Mongols)—the former was actually appointed by the Pope as archbishop there. Late in the thirteenth century, a traveler from Peking journeyed west to Europe: Rabban Sauma was a Nestorian monk, who set out on a pilgrimage to Jerusalem. In fact, he went further—he visited Constantinople, talked with university students in Paris, and met King Edward I of England, who in return sent envoys to the Mongol ruler of Persia.

In 1340, Francesco Pegolotti, agent for a Florentine banking family, prepared a handbook for merchant travelers that points to a flourishing trade with Asia. These links were broken when the Mongol empire disintegrated during the fourteenth century. By the fifteenth century the possibilities of European trading with Asia were being blocked even further by the creation of the Turkish Ottoman empire, which eventually controlled not only the area now known as Turkey, but the whole eastern and southern Mediterranean coasts.

The medieval travelers had brought back much valuable geographical knowledge about Asia—incorporated, for example, in the *Catalan Atlas* of 1375, made for the king of Aragon by Abraham Cresques. By the early fifteenth century numerous Latin manuscripts of Ptolemy's *Geography* were circulating in western Europe, and though his maps were inaccurate, his methodology stimulated much improvement in cartography. New information about South Asia was provided by the early fifteenth-century Venetian merchant traveler Nicolo de Conti, who had reached the East Indies and speculated on the possibility of reaching the Orient by sailing round Africa.

The first people to be inspired by these developments were the Portuguese. Their prince, Henry "the Navigator," inaugurated at Sagres a national program for exploring the world, making the place a center for the development of cartography, navigation and shipbuilding. Jehuda Cresques (son of Abraham) was brought there to piece together the facts brought back by the seafarers. Mariners were instructed in the keeping of accurate logs and charts. The latest navigating instruments were tested, and new types of ship were constructed. One expedition after another was sent out to explore the West African coast, and decade by decade they pushed on further. Dinis

Dias rounded the western tip of Africa in 1445, Bartholomeu Dias the southern tip in 1487. Ten years later, Vasco da Gama set out from Lisbon, rounded Africa, sailed up the east coast and across the Arabian sea to Calicut, on the southwest coast of India. The Portuguese established viceroys in India, who extended their control throughout the Indian ocean and opened up trade with China.

Meanwhile, Cristoforo Colombo from Genoa had been campaigning for support to sail west from Europe to reach China. According to his calculations (based on a low estimate of the size of the earth made by Marinus of Tyre in A.D. 100) the distance was much less than the Portuguese had already sailed down the West African coast. He eventually convinced King John of Portugal (Henry's son), and in 1492 took his ships across the Atlantic to reach the landmass we now call the Americas. Both the Portuguese and the Spanish then began to explore and conquer Central and South America. In North America, the Spanish were the first to establish a colony, in Florida, 1565. An English settlement was created at Jamestown in 1607, a French at Quebec in 1608, and a Dutch at New York in 1625.

In 1519, Ferdinand Magellan (a Portuguese who at first worked for King John) on behalf of Spain set out to try to round the southern tip of America. He succeeded, sailed north along the coast, and then across the Pacific to the Philippines. Here he was killed, but one of his ships continued south and west, rounded Africa and at last arrived back in Seville—18 shipworn men out of the original crew of 250. In principle, the whole world was now open to European penetration, although much remained to be explored and discovered during the following centuries.

By 1700, an international trading community had come into being that ran round the globe. For example, the Portuguese had trading stations or possessions in such places as Angola, Mozambique, Goa in West India, Macao in China, and Brazil, and had made contact with Japan; Spain controlled most of South and Central America, Florida and the southwest of North America; Holland was in control of Ceylon, South Africa, and the East Indies; France had footholds in Mauritius, Pondicherry in southeast India, and Canada. Britain controlled parts of the African coast, a few coastal areas of India, and most of North America.

The peoples who now felt the impact of the Europeans were at very varied levels of development. Some, such as many of the North American Indians and African tribes, still had a hunting and gathering economy. On the other hand, both China and India were civilizations of great antiquity, with cultures no way inferior to the European late Middle Ages. They had their own traditions of scholarship, their own contributions to technology, mathematics, astronomy, medicine, and chemistry, which we have not been able to reflect in this book. The Europeans who came to trade and administer settlements in these countries paid little attention to their cultures, though something was learned about traditional Indian herbal medicine.

Thus the knowledge that existed outside Europe made no direct contribution to the continued development of science. But European sailors and traders brought back reports of new animals and plants, and introduced new crops such as tobacco and the potato. In the sixteenth century there appeared accounts such as the *Coloquios dos simples e drogas he cousas medicinais da India*, by the Portuguese Garcia d'Orta, the *History of the Indies*, by Oviedo y Valdes, and *Joyefull Newes out of the Newefound World*, by Nicholas Monarde, which provided much new information about plants and animals. Zoos were established in Europe, one of the first being at Dresden in 1554. Botanical gardens and arboreta were formed that included plants brought back by the explorers. For example, René du Bellay and Pierre Belon in France planted seeds from western Asia, and Charles de Lécluse (Clusius) brought together a collection of flowering bulbs at Leyden. As the discovery of new plants gained impetus, more and more botanic gardens were created as repositories of the world's plants. In England, the private seventeenth-century collection of Henry Capel became in 1759 what is now the Royal Botanic Gardens at Kew. The discoveries and collections drove home the point that naturalists would eventually have to provide descriptions of an ever widening array of new species. In the eighteenth century, naturalists would participate in new voyages of exploration (for example, Joseph Banks on James Cook's first voyage to the Pacific Ocean and Australasia).

New scientific traditions, based on European science, would eventually be created in many countries outside Europe—for example, the Americas, India, Japan—and later we shall be paying attention to these developments and their implications for communication. But for the present, we return to a consideration of the situation in eighteenth-century Europe.

Section 5

The Eighteenth Century

THE EIGHTEENTH CENTURY

European science during the first half of the eighteenth century paused somewhat after the excitement of its formation as a profession and the achievements of seventeenth-century giants such as Isaac Newton. But it soon spread to many new centers within Europe, and many more societies were formed. Science was stimulated by the growing pace of industrial development. Roads (and the postal service) and shipping improved. The use of coal much increased. Cast iron came to be more readily produced, and this led to the formation of large industrial undertakings such as those of Abraham Darby at Coalbrookdale and the Carron Ironworks established by John Roebuck.

After the mid-century the pace of development was such that the ensuing period is commonly known as the "industrial revolution." It was triggered by a great increase in the market (both in Europe and overseas) for textiles, primarily cotton. The raw material was, of course, imported into England from its colonial possessions, such as America and India. The developments that took place in Europe consisted in the mechanization of the processes of textile manufacture, such as weaving and spinning. Inventions such as the spinning jenny of James Hargreaves (1764), the waterframe of Robert Arkwright (1769), the mule of Samuel Crompton (1779), and the power loom of Edmund Cartwright (1785) made possible a great increase in output. Parallel to this was the development of the steam engine—at first for pumping mines (Thomas Newcomen, 1712), then as a machine with many applications (James Watt, 1765)—and this gave the textile industry an enhanced source of power.

The men responsible for these inventions were mainly skilled craftsmen, although Watt's work did have as a background the quantitative studies on heat of the scientist Joseph Black, for whom he had worked as an instrument maker. Many of the scientists themselves were amateurs. In the eighteenth century (up to 1785), according to Inkster, in the United Kingdom only 55 percent of scientists and 32 percent of engineers had any university connections. During the industrial revolution, there was much coming and going between laboratory and workshop. Inventors, industrialists and entrepreneurs came from every social class, particularly the clergy and doctors, as well as artisans.

One way in which scholarly learning was transmitted to craftsmen eager to learn was through the publication of technical encyclopedias. Early examples are the *Dictionnaire des arts et de sciences* of Thomas Corneille (1694) and the *Lexicon technicum* of John Harris (1704). Ephraim Chambers in 1728 published the *Cyclopaedia or Universal Dictionary of Arts and Sciences*, which through various editions was in use throughout the century. In the 1730s came the start of Johann Zedler's *Universal Lexicon* in sixty-four volumes. Young Denis Diderot undertook to translate Chambers' work into French, but instead in 1751 began to publish the collectively prepared *L'Encyclopédie ou Dictionnaire raisonné des sciences, des arts et des métiers*, perhaps the most powerful intellectual force of the century, widely purchased and read outside France. The first issues of the *Encyclopaedia Britannica* began to appear in 1768, edited by William Smellie. Lastly may be mentioned the vast uncompleted enterprise of Charles Panckoucke, the *Encyclopédie méthodique*, the first of whose 166 published volumes appeared in 1781.

THE BURGEONING OF SOCIETIES

The eighteenth century witnessed a dispersal of science throughout Europe. Before 1750, academies had been set up in Berlin, St. Petersburg, Stockholm and a number of other cities. From mid-century, the pace of formation quickened. In France, between 1700 and 1776 about one hundred academies were formed, including centers such as Bordeaux, Rouen, Dijon and Lille. A provincial academy was formed in each of the German states. During the century, eleven towns in Italy formed scientific academies, including Bologna and Turin.

In these societies scholars were actively engaged in collecting information, surveying and planning experimental programs. Most of the academies published proceedings, and nearly all of these devoted their pages to summarization, translation and popularization of the advances in knowledge. Science had begun to split up into a number of sciences, each with its special interests and methods, each becoming less comprehensible both to laymen and to specialists in other branches.

This specialization was reflected most strongly in the type of society formed in England. The earliest specialist societies seem to have been those devoted to natural history—for one thing, this subject was eminently suited for study by the growing army of doctors, schoolmasters, clergymen and manufacturers with an interest in science. A short-lived botanical society was founded in London in 1721 by Johann Dillen and John Martyn. The Aurelian Society for entomology was formed in 1745—the Society of Entomologists was not formed till 1780. The year 1782 saw the founding of the Society for Promoting Natural History, and some of its members later formed the Linnean Society. This was in 1788, the year that the great Swedish botanist Carl Linnaeus died. His herbarium, library and manuscripts were purchased for the new society.

Short-lived chemical societies were formed in Edinburgh (1785), Glasgow (1786) and London (1794). Mathematics was early represented by the Spitalsfield Mathematical Society formed in 1717, which seems to have been composed of local tradesmen and professional men. It lasted till 1846, when it was absorbed by the Royal Astronomical Society. In 1771 the engineer John Smeaton gathered together in London a group of like-minded men to discuss their work, forming a Society of Civil Engineers. Numerous agricultural societies were set up to popularize scientific agriculture—e.g., the Bath and West of England Society (1777), the Highland Society (1784), and the Smithfield Club (1798).

Of particular importance in England in the eighteenth century were the provincial societies that came into being under the stimulus of growing industrial development. The Royal Society had lost its early interest in trades and manufactures. The two leading industrial towns of the period were Manchester and Birmingham, and it was in them that manufacturers and scientists came together in societies that made a strong impression on English intellectual life.

In Manchester a club that met weekly to discuss literature and science became in 1785 the Literary and Philosophical Society. Chemistry was a strong interest of the society in its early years, its most famous scientific member being the chemist John Dalton. The Birmingham group was the

Lunar Society, founded about 1766 by the engineer Matthew Boulton. Its membership included Boulton's partner, James Watt, Erasmus Darwin (grandfather of Charles), the chemist Joseph Priestley, and the industrialist Joseph Wedgwood. As was said of another Philosophical Society founded at that time in Hull, "from this establishment more learning has been diffused throughout the town than could ever have found its way along the private channels of a few rich individuals."

A society of a different kind was founded in the middle of the century, the Society for the Encouragement of Arts, Sciences and Commerce, now known as the Royal Society of Arts. William Shipley, a Northampton drawing-master, saw the need for the rapid development of industry and agriculture to overcome poverty, and conceived of a society that would offer prizes to inventors and would indicate to would-be inventors the areas in which improvements were most needed. He succeeded in interesting a number of influential people, who formed the society in 1754. It was a success. So many inventions and suggestions came in that subject committees had to be set up—agriculture, chemistry, fine arts, manufactures, mechanics and trade. The proceedings of the society were published in various places, a series of annual transactions beginning in 1783, which were replaced by the *Journal of the Society of Arts* in 1852.

The society set up a system of honorary corresponding members, in Russia, Germany, Sweden, Switzerland, France and Italy. Its reputation led to the founding of similar bodies in Europe, such as the Patriotische Gesellschaft of Hamburg (1763), the Free Oeconomical Society of St. Petersburg (1766), and the Société pour l'Encouragement de l'Industrie Nationale in Paris (1801).

Another important link between science and technology was the establishment of mining "academies" (in the educational sense) to train mining engineers and administrators in the German states. Already from 1560, state positions for chemistry and mineralogy had been established at the Dresden Museum for Mineralogy and Geology. From 1700 on, state support for mining education was provided at Freiburg, and a formal academy was set up in 1765. Similar institutions were later founded in Schemnitz, Prague, Berlin and St. Petersburg, to say nothing of the Ecole des Mines in Paris. The society academies brought mutual encouragement to their members. Their prize competitions and collaborative projects stimulated scientific development. They began to create new career opportunities for scientists, and to provide that system of peer assessment that is essential to the ethos of science.

MUSEUMS OF NATURAL HISTORY

The eighteenth century saw the birth in Britain of a national institution that has played an important part in intellectual life—namely, the British Museum (including its natural history side), which originated in the collections of Hans Sloane. He was president of the Royal College of Physicians, 1719–35. Apart from the publication of a natural history of Jamaica, his energy and money was devoted to the organization of science and to the building up of what was at the time an unsurpassed collection of specimens (botanical, zoological, geological) and antiquities.

The collection was made to illustrate specific scientific topics, and to serve as reference and research material. In his own words:

> It were therefore much to be wished for and endeavored that there might be made and kept in some Repository as full and complete a collection of all varieties of natural bodies as could be obtained, where an enquirer might be able to have recourse, where he might peruse and turn over and spell and read the book of nature, and observe the orthography, etymology, syntax and prosody of nature's grammar, and by which, as with a dictionary, he might readily turn to find the true figure, composition, derivation and use of the characters, words, phrases and sentences of nature, written with indelible and most exact and most expressive letters, without which book it will be very difficult to be thoroughly a literatus in the language and sense of nature. The use of such a collection is not for divertisement and wonder and gazing, as 'tis for the most part thought and esteemed, and like pictures for children to admire and be pleased with, but for the most serious and diligent study by the most able and proficient in natural philosophy.

Sloane worked towards such a repository, acquiring collections made by eminent botanists, zoologists, geologists and antiquarians. He also collected books—his library had reached a total of forty thousand volumes and four thousand manuscripts at his death. His private museum acquired such a reputation that scientists looking for a safe home for valuable books, papers or specimens automatically thought of donating them to him. In his will, he directed that his trustees should endeavor to have his collection taken into national control. Eventually, the manuscript collections of Robert Cotton and Robert Harley were added to the Sloane collections and formed the nucleus of the British Museum that was founded in 1753.

The artifacts and books of the British Museum collections for some time overshadowed the natural history collection, and it was not until 1880 that the anatomist Richard Owen was able to supervise the transfer of the collection to its own building in South Kensington, London. In the interim, the Paris Museum d'Histoire Naturelle, created in 1790 from the old French royal gardens, had achieved a great reputation as a center for teaching and research, with twelve professorships and highly trained assistants. Its example did much to stimulate the call in Britain for improvement of natural history facilities.

SPECIALIZED JOURNALS

By the end of the seventeenth century, the scientific journal was firmly established. Barnes notes that over three hundred learned journals were founded in the period 1665–1730. During the first half of the following century new academy publications appeared at Berlin (1710), St. Petersburg (1725), Stockholm (1739) and Göttingen (1739), and new learned journals at Venice (1728), the Hague (1737) and Hamburg (1747). In 1752, the first volume of *Commentarii de rebus in scientia naturali et medicina gestis* was published at Leipzig: this useful and successful journal, which continued until 1798, gave news of the work of the academies and details of new books. Its book reviewing was imitated in England by the *Medical Commentaries* (1773–95), a quarterly publication containing an account of the best new books in medicine, medical cases and observations, medical news, and a list of new medical publications. Its editors observed that few had the leisure to read all that was new: "a scheme, calculated for saving time in reading, and expense in purchasing books, is a concise view of the books themselves."

In the 1770s a significant new development occurred in scientific journalism—the appearance of specialist periodicals. The reasons for this development were set out in the first issue of the *Observations sur la physique, sur l'histoire naturelle, et sur les arts* (Paris 1771): first, the academy publications were only printed some considerable time (perhaps several years) after a paper was read to the academy; second, the academy proceedings were in various languages. As a result, it was stated, scholars remained ignorant of each other's work, and wasted precious time working on problems that others had already solved. With the growth of national academies, and the increase in the number of scientists within each country, correspondence between nations had slackened. What was needed was a new organ of communication that was both more specialized and international.

The *Observations* were edited during 1771–79 by François Rozier of Lyons, priest and farmer, and continued publication until 1793. Although largely concerned with physics, it dealt also with chemistry and other sciences. More specifically chemical were the journals edited by Lorenz Crell from 1778 onwards: the *Chemisches Journal* (1778–81), *Die neuesten Entdeckungen in der Chemie* (1781–86), the *Chemisches Archiv* (1783–91), and the *Chemisches Annalen* (1784–1803). The *Taschenbuch für Scheidekünstler und Apotheker*, founded by Göttling, was published 1780–1829. In 1787, William Curtis founded the *Botanical magazine*, designed to illustrate and describe the most ornamental cultivated foreign plants. Each issue contained colored plates, some drawn by James Sowerby but most (until 1815) by Sydenham Edwards.

The most important chemical journal of the period was the *Annales de Chimie*, edited by de Morveau, Lavoisier and the rest of the brilliant group of French chemists who developed modern chemical nomenclature. The first volume, 1789, repeated Rozier's complaint that new discoveries were scattered in publications in many different languages. The editors proposed to present in the journal accounts of their own work, memoirs from foreign correspondents, and notices of papers presented to foreign academies. The first volume included papers presented to the Académie des Sciences, correspondence between Crell and the French chemists, translated extracts of articles in the *Chemisches Annalen*, and book reviews.

In 1797, William Nicholson launched a *Journal of Natural Philosophy, Chemistry and the Arts*. In part an imitation of the Continental journals already mentioned, it was the first English journal of this type to be published independently of the scientific societies. Nicholson was able to offer far speedier publication than did the society journals. In his time, papers in the *Philosophical Transactions* might not appear for a year after being delivered, and in the case of the Manchester Literary and Philosophical Society, papers might not see print for five years. In *Nicholson's Journal*, as it became known, papers often appeared within a fortnight of delivery. In 1813, it was absorbed into the *Philosophical Magazine*, edited by Alexander Tilloch, which had been published in London since 1798. In the physical sciences this journal soon became second in importance to the *Philosophical Transactions.*

The titles mentioned give only a sampling of the periodical output of the eighteenth century—Bolton lists sixty-nine new chemical journals originating during 1750–99, apart from the publications of the academies—but they do give some idea of the new developments. Over the whole period 1665–

Table 4 Scientific Periodicals, 1665–1790

Type	*1665–1699*	*1700–1749*	*1750–1790*	*Total*
substantive	20	59	422	501
proceedings	8	47	209	264
collections	–	7	74	81
dissertations	–	9	31	40
abstracts	–	6	36	42
reviews	2	2	36	40
almanacs	2	1	44	47
others	3	10	24	37
total	35	141	876	1,052

Source: Kronick (1976)

Table 5 Subject Spread of Periodicals

Type	*Gen*	*Med*	*Biol*	*Phys*	*Tech*	*Agric*	*Other*
substantive	–	186	151	20	49	28	59
proceedings	178	34	–	3	6	37	9
collections	44	63	–	–	–	–	13
abstracts	36	36	–	–	–	–	10

Source: Kronick (1976)

1790, Kronick identified 1,052 scientific periodicals and analyzed them in table 4.

Some of these types he analyzed by subject, as in table 5.

Abstracts were of considerable importance, many of them being published in Germany, providing discrete information in condensed and simplified form. As specialization increased towards the close of the century, there was a lessening of the "popular" appeal of journals (as had, for example, been stressed by de Sallo). Indeed, Rozier sternly disavowed it: "We are not offering pleasant reading matter to leisured amateurs, nor the sweet illusion that they are being initiated into sciences of which they know nothing." Science was now an organized profession, and its channels of communication had to be kept free of all trivial, stale or ill-digested matter.

THE PROGRESS OF BIBLIOGRAPHY

Specialist bibliographies continued to appear in the eighteenth century—for example, the *Prodromus bibliothecae metallicae* of Jacob Leupold

(1700); the *Bibliotheca scriptorum historiae naturali* of J.J. Scheuchzer (1716); and a catalogue of alchemical books in the *Histoire de la philosophie hermétique* of Nicolas Lenglet du Fresnoy (1742). The indexing of the multiplicity of magazine articles, first attempted by Cornelius à Beughem, was taken up, for medical papers, by Michael Alberti. His *Tentamen lexici realis observationum medicarum* (1727–31) was an alphabetical subject index to seventy-six sources, including the periodicals *Acta naturae curiosorum*, *Acta eruditorum lipsiensium*, *Breslauische geschichte* and *Acta Berolinensia*. The two volumes included over forty thousand entries. The great systematist Carl Linnaeus in 1736 published a *Bibliotheca botanica*, and also edited the *Bibliotheca et philosophia ichthyologica* of his friend Petrus Artedi (1738). The botanical work was in systematic order; the authors of its one thousand books were classified by Linnaeus as in table 6.

In mid-century other fields of science were surveyed, e.g., by J.F. Weidler's *Bibliographia astronomica* (1755) and J.E. Scheibel's *Einleitung zur mathematischen Bücherkenntnis* (1769 on).

In the 1770s, scientific bibliography was dominated by Albrecht von Haller of Switzerland, who was perhaps the most prolific and versatile

Table 6 Classification of Botanical Authors

- Botanici
  - Collectores
    - Patres (Greek, Roman, Asiatic, Arab, Medieval)
    - Commentatores (on the Patres)
    - Ichniographi (illustrators)
    - Descriptores
    - Monographi (arranged by plant genus)
    - Curiosi (rare plants)
    - Adonistae (catalogues of botanical gardens)
    - Floristae (flora, by country)
    - Peregrinatores (travels)
  - Methodici
    - Philosophi
    - Systematici
    - Nomenclatores
- Botanophili
    - Anatomici
    - Hortulani (vegetable culture)
    - Medici (herbals)
    - Anomali (poets, biographers, bibliographers)

Source: Linnaeus

scientific writer of all time, responsible for thirteen thousand scientific papers, in botany, human physiology and anatomy. The final volume of his *Elementa physiologiae corporis humani* (1757–82) included a list of the works he had consulted, itself over one hundred pages. His first purely bibliographical publication was the *Bibliotheca botanica* (1771–72), which listed fifteen thousand books and papers. He followed this with bibliographies of surgery (1774–75) and anatomy (1774–77), and then four volumes of a *Bibliotheca medicinae practicae* (1776–88), with fifty-two thousand references. Haller's bibliographies were divided into chronological sections, listing authors alphabetically within each section, giving details of their lives and writings, with notes on the books cited. Perhaps, wrote Brodman, "Haller can be considered the last person who was able to know the entire literature of medicine. Later medical bibliographers were to be faced with the necessity of limiting their work in one or more of several alternative ways: by period covered, by language, by subdivision of subject, by country of origin, or by some other means; or else they were to be forced to assemble teams of assistants to aid them."

In the decades that followed Haller's work, the number of scientific bibliographies published increased considerably, and we cannot attempt to list them. An important subject catalogue of medicine was the *Bibliotheca medico-practica et chirurgica* of W.G. Ploucquet, published in series over the period 1793–1813, containing several hundred thousand references to books, dissertations and articles. Ploucquet faced the alternatives of cumulation (and hence the cost of reprinting) or supplements (and hence inconvenience in consultation), which every bibliography faces if it is to remain up-to-date. He discussed other problems of the bibliographer, as Brodman has summarized:

1. The tremendous growth of literature. "The job would be simpler if the legacy were smaller, but the wealth of material overwhelms us, and we are blinded by too much light. . . . To make matters worse, no day passes but someone throws another article upon this mountain of material. . . . Our life is too short, and there are so many books; money is so scarce, and there is so little time."
2. It is impossible to tell the subject of a book or article from the title alone, and reading it takes time. "It is obviously insufficient to record only the titles. . . . Titles often promise more than they deliver, sometimes less, sometimes matter of which the title gives no inkling."
3. Many things important to medicine are found in nonmedical works. "Valuable material is often included in histories, travel diaries, and

in various genres, where it is least expected." This expands the work enormously, and makes it more difficult to complete. "Many will say that priority should perhaps have been given to those who wrote about disease, since their work offers the most return for the least effort."

4. Many of the writings indexed are worthless from the scientific point of view, yet they must be included. "A compiler cannot afford to indulge in the arrogance of deciding what is beneath notice. . . . Besides, the profession's favour may change, and what has previously been condemned may later be approved. . . . My object is after all not to be critical. It is the recording, so far as possible, of all that has been done, said, seen, observed by physicians and others, of all ages and nations—whether right or no."
5. It is impossible to see all the works to which references are made. "Many works are known to me only by title, and although I have read many completely, judgment on others could not be made. . . . It must be admitted that if I had at hand the originals of much of the material, or if I could have obtained the best editions, the work would have been the better for it. But it is doubtful that even the best of libraries could possess itself of such a treasure."

Such were Ploucquet's problems that continue to plague the bibliographer today.

TECHNICAL TERMINOLOGY

Almost every step in the progress of science has been marked by the formation or appropriation of new technical terms. Common language has, in most cases, a certain degree of looseness and ambiguity, reflecting the vagueness and indistinctness of common knowledge. As knowledge became more exact, terms were needed that would convey meaning steadily fixed and rigorously limited. When knowledge still further advanced, terms—as well as being precisely defined—had to be systematically related to each other.

Any such system comprised two groups of terms: (1) the nomenclature or system of names given to the natural objects studied—minerals, chemicals, plants, animals, atoms, stars or galaxies; and (2) the terminology, or technical terms used to describe their parts and properties, and the processes in which they take part. These two sets of terms went hand in hand. The terminology was indispensably necessary to give fixity to the nomenclature:

for example, the recognition and naming of the kinds of plant depended on the exact comparison of their resemblances and differences—features which, to be readily and widely communicated, had to be expressed in agreed technical terms.

The decisive advances in scientific nomenclature—primarily in biology and chemistry—were made in the eighteenth century, associated with the names of Linnaeus and Lavoisier. Each of these men, building on the work of predecessors and with the help of collaborators and successors, founded systematic nomenclatures for their sciences.

First let us glance at the general way in which scientific terminology and nomenclature developed. In the early stages of a science, technical terms were taken from words in common usage, by rigorously defining or otherwise fixing their meaning. In this way, almost all the fundamental terms of geometry were formulated by the Greeks, such as in table 7.

This same process continued throughout the centuries in many fields of science, although with decreasing frequency. Thus "al qali" (alkali) at first in Arabic denoted the ashes of a particular plant, but later came to be applied to all earths with certain chemical properties. Again, in the eighteenth century, the word "corpuscle" (a small body) was applied to cells in mammalian blood.

Words thus borrowed from common language had the advantage that they were understood after a short explanation, and were retained readily in the memory. But users not in the special field in which they originated were liable to employ them with insufficient precision. Consequently, scientists soon began to make use of a second class of terms, specially constructed or imported, and perhaps containing some indication or description of the concept that they represented.

The translation of foreign texts often involved the importation of new terms of this kind. For example, the medieval astronomers adopted Arabic words such as zenith, nadir and azimuth when making their Latin translations, and such words passed into modern languages. Similarly from the Arabic came such chemical terms as naphtha, talc and tartar, together with

Table 7 Geometrical Terms

technical term	*original Greek usage*
sphere	handball
cone	spinning top
trapezium	table with one side shorter than the opposite

Persian words (borax, cinnabar, gypsum, laudanum) and some from Sanskrit (camphor, indigo, sugar, sulphur). Botanical terms from the Arabic included artichoke, coffee, lilac, jasmine, saffron.

There were many examples of the importation by translation of words into English: King Alfred introduced into Anglo-Saxon the terms arithmetic, astronomy and mechanic, and even before his time the words acid, cancer, circle and comet had come into the language from Latin. In 1542 Robert Copland translated extracts from the fourteenth-century Guy de Chauliac's *Grande chirurgie*, and introduced into English such words as abdomen, cartilage, optic and tendon.

For modern European languages, the major sources of borrowed words were Latin and Greek. From the Latin came such terms as apex, axis, cirrus, corona, cortex, focus, humus, latex, locus, saliva, species, stamen, tuber, viscera. Early Greek importations into English were at first indirect. In the period before 1500, Greek words that had been adopted first into Latin and then into French were introduced into English, e.g., atom, center, dropsy, galaxy, horizon, magnet, sphere and turpentine. Others came directly from Latin: artery, cycle, dactyl, iris, meteor, pole, phial and thorax before 1500, and in the sixteenth century chord, cylinder, nausea, larynx and prism. Direct importations from Greek came later, for example, cardiac, cytology, embryo, gastritis, hormone, mycelium, pectin, pepsin, petroleum and rheology. Often, what had been a common word in Latin or Greek was adopted in modern languages with modified meaning as a technical term, as in table 8.

The names of plants, minerals and other natural objects were often taken from the places where they occurred conspicuously, or from persons associated with them. So from places were named French marigolds and Epsom salts; from people, fuchsias, dahlias and Glauber salts, as also Halley's

Table 8 Technical Terms and Their Roots

technical term	*meaning of root*
amnion	bowl
bacillus	little stick
bacterium	little stick
cotyledon	cup-shaped
electricity	amber
larva	mask
nucleus	nut
toxin	arrow

comet. (Later, during the nineteenth century, many physical units were named after famous scientists—such as the ampere, the coulomb, the farad, the gauss, the henry, the joule, the ohm, the volt and the watt).

The myriad technical terms of science were therefore drawn from the most diverse sources. Many of them, despite their casual or fanciful origin, became sufficiently familiar and fixed in meaning to serve the needs of science. But this was not always the case. At times, there accumulated a great number of terms—for example, the names of minerals, chemicals, plants or animals—which had been assigned haphazardly throughout the ages, by people of many different tongues, under the influence of many different theories. When this occurred, the chaotic nomenclature obscured whatever real relations had been shown to exist among the species named. A reform became necessary—the creation of a new nomenclature and terminology based upon a systematic classification of all the known species. The most dramatic of these reforms took place in the eighteenth century in biology and chemistry, and the course of development in each case will be traced.

The technique of classification was used in the following way to aid in the identification of the many known species of plants, animals and substances. The terms, by which significant characteristics of the species were described, were first carefully defined; these terms were then used to group the species into a hierarchical system; from this system their names were then derived. Thus terminology, classification and nomenclature developed hand in hand.

The choice of concepts—and hence terms—by which to differentiate biological or chemical species was often guided by an intuitive feeling that some groupings displayed the "natural affinity" of species better than other groupings, which were held to be "artificial"—a feeling engendered by repeated observation of the species studied. Nevertheless, the search for a natural system, based upon an accepted "generative principle," often had to give way to a system acknowledged to be artificial but convenient in use.

THE LANGUAGE OF BOTANY

Even among uncivilized, illiterate peoples, many hundreds of plant species were known and named. In prehistoric times, a great variety of plants was already important to man—green vegetables such as cabbage, lettuce, spinach, nettle, cress, beans and peas; carrots and radishes; fruits such as apple, pear, plum and cherry; walnuts and several oil seeds; cereals such as wheat, barley, rye, oats, millet, and rice; mustard and cane sugar. Gourds were in use as vessels; flax, hemp and cotton as fiber plants; dyes were obtained from madder, saffron, woad and indigo. An early knowledge of plants can be deduced from comparative studies that show the existence of several plant names common to all the Indo-European languages, which started to diverge at least four thousand years ago. For example, the beech tree is *fagus* in Latin, *phegos* in Greek, *bok* in Old Slav, and *buohha* in Old High German; allowing for well-known soundshifts, these words are cognate. Thieme suggests common roots for other tree names, cited in decreasing certainty of identity: birch, aspen, oak, yew, willow, spruce, alder, ash. There is also a root for "cereal," which is sometimes applied to barley and sometimes to cereals in general. (To stray for a moment into zoology, Thieme's identifications for animal names common to Indo-European languages are: wolf, bear, lynx, eagle, falcon, owl, crane, thrush, goose, duck, turtle, salmon, otter, beaver, fly, hornet, wasp, bee, louse, flea, dog, cattle, sheep, pig, goat.)

The common names that all peoples have given to plants usually in some measure describe the plants or suggest something or someone connected with them. The name might refer to the plant's appearance (buttercup, silverweed), to its taste or smell (roastbeef plant, sauce-alone), to its habitat (watercress), to its behavior (chokecherry, catchfly), or to its use (goutweed). Over a wide area, a plant might be called by different names, such as pennycress, fanweed, stinkweed; or the same name (bluebell) might be applied to several plants. Differences in usage in time became highly confusing when botanists from different regions tried to exchange experiences.

The earliest known writer to discuss the grouping and naming of plants was Theophrastus in the fourth century B.C. In his *History of Plants*, he recognized five hundred species, and tried to group them into classes: trees, shrubs, undershrubs and herbs, with further distinctions such as cultivated and wild, flowering and flowerless, fruit-bearing and fruitless, deciduous and evergreen, spiny and spineless, or terrestrial, marshy and aquatic. The groupings in the *Materia medica* of Dioscorides (first century A.D.) were usually based on uses—aromatic, alimentary, medicinal, vinous. In the thirteenth century, Albertus Magnus advanced systematic botany by comparative studies of the roots, stems, leaves, flowers, fruits, bark, pith and other parts of plants.

The first Renaissance writer in whom there were traces of plant grouping was Hieronymus Tragus (Jerome Bock), who published an herbal, the *Kreutterbuch*, in 1539. His descriptions of plants were simple and homely. His classification of plants was still largely based on their uses, but he also paid attention to the shapes of leaves, the ramification of branches and roots, and the size and color of flowers. There were also sixteenth-century contributions to botanical terminology: the German Leonhard Fuchs in his *Historia stirpium* (1542) gave a four-page glossary of terms such as fruit, berries and stamen; later, the *Stirpium historiae pemptades* of the Belgian Rembert Dodoens (1583) contained apt definitions of the words root, leaf, flower, calyx, anthers and catkin.

The French naturalist Charles de l'Ecluse (Clusius), in *Rariorum plantarum historia* (1576), grouped together (1) trees, shrubs and undershrubs, (2) bulbous plants, (3) sweet-smelling flowers, (4) odorless flowers, (5) poisonous, narcotic and acrid plants, (6) plants with a milky juice, (7) others—a scheme mostly based on uses. He introduced more exactness into plant description, as later did the Swiss Kaspar Bauhin in his *Prodromus theatri botanici* (1620), who noted all the obvious parts of the plant in a fixed order: the form of the root, the height and form of the stem, characteristics of the leaves, flowers, fruit and seed. Bauhin also constructed a large synonymy of plant names, covering six thousand plants, in his *Pinax theatri botanici* (1623), and systematically used a binomial nomenclature, such as *Ranunculus montana* for mountain crowfoot.

All the Renaissance botanists so far mentioned worked in northwest Europe. Meanwhile, in Italy Andrea Cesalpino was laying the foundations of a more systematic approach. In *De plantis* (1583), he wrote: "In the immense multitude of plants, I see that want which is most felt in any other unordered crowd: if such an assemblage be not arranged into brigades like an army, all must be tumult and fluctuation." He therefore attempted to make such an arrangement for the whole range of plants. He showed that whether we use the roots, the stems, the leaves or the blossoms to group plants, we separate species that are alike in other ways. Using criteria such as the nature of the stem (woody or herbaceous) and then the number, position and form of seeds, conceptacles and fruits, he developed a methodical classification, introducing and defining many new terms.

During the seventeenth century, the number of known plant species increased rapidly, as a result of expeditions in the Americas, Asia and Africa. The number of botanical gardens grew. The intensive study of the new material stimulated the development of even more carefully designed technical terms for the parts and processes of plants. This was particularly due to Joachim Jung in his *Isagoge phytoscopica* (1678)—much of his terminology has survived till today. In the period 1686–1704, the English botanist John Ray published his *Historia plantarum*, in which he described nineteen thousand plants. He included Jung's *Isagoge*, adding comments of his own.

He stressed that to achieve a more natural classification, all the characters of a plant should be taken into consideration, not just a few.

THE LINNEAN SYSTEM

Despite the vast amount of work on plant description done by the end of the seventeenth century, there was still a need for a good, comprehensive working classification, the coordination of all synonyms for plant species, and the establishment of a convenient system of nomenclature. The completion of these tasks was virtually the accomplishment of one man, the Swede Carl Linnaeus.

His first work was published in 1735, at the age of twenty-eight: the *Systema naturae*, a comprehensive sketch of the whole domain of natural history. A series of botanical works followed in the next twenty years, later appearing in successively revised editions. The great talent of Linnaeus was systematization: he might almost be said to have been a classifying, coordinating and subordinating machine. He brought together all three aspects of botanical taxonomy: terminology, classification and nomenclature. He adopted and extended Jung's *Isagoge* with Ray's supplements, providing a glossary of many hundreds of technical terms for plant description.

His classification was rigidly and consistently founded on the sexual organs of plants, the stamens and pistils. The function of these plant organs had been discovered in the seventeenth century by the English scientist Nehemiah Grew and the German Rudolphus Camerarius, and Linnaeus arranged all plants according to their number and disposition (plants with no apparent organs of reproduction he called the cryptogams). Using these sexual characteristics, he divided plants into twenty-three classes, and each class further into orders. The subdivision of orders into genera was based on characteristics of the fruit and flower, and the division of genera into species on still more varied characteristics. His scheme, though acknowledged to be "artificial," proved to be very easy for botanical use.

A further feature contributing to the success of the Linnean system was his consistent use of a binomial nomenclature. Despite the earlier use by Bauhin, species were often still designated by adding a descriptive phrase to a generic name—e.g., the sweet briar rose was known as *Rosa aculeata, foliis odoratis subtus rubiginosis* (prickly rose, having odorous leaves with red-brown underside). Linnaeus himself defined the species as "rose, germs globular and peduncles smooth, stem with prickles scattered straight, petioles rugged, leaflets acute." A name could be created using these characteristics in succession, thus *Rosa globosa glabra scabra acuta*, but this systematic procedure would lead to long names. Linnaeus introduced the use of "trivial" specific names—a single conventional word, imposed without any general rule. The sweet briar became *Rosa eglanteria*.

This practice was soon accepted as standard, and its development was eventually guided by a set of *International Rules of Botanical Nomenclature*. These rules were first drawn up by the French taxonomist Alphonse de Candolle, and agreed by an international congress in 1867. They have been amended over the years. For most plant groups, the names given by Linnaeus were accepted as standard. The rules stipulated that generic names such as *Rosa* might be taken from any source whatever; specific epithets such as *eglanteria* should give some indication of the appearance, characters, origin, history or properties of the species.

KNOWLEDGE OF SUBSTANCES

Even in ancient times, man had knowledge of an amazing variety of minerals. The use of metals such as copper, gold, silver, tin, iron, lead and mercury presupposed a knowledge of their ores and their oxides, as well as other derivatives such as the vitriols (sulphates). Clay, chalk, lime, gypsum, and other useful minerals such as talc, emery and asbestos; common salt, soda, potash and alum; sulfur, soot, charcoal and diamonds—these and many other inorganic materials were known to antiquity. Painters used ochre, cinnabar, orpiment, white lead, red lead and other coloring matters. The range of known organic substances was also wide—petroleum, asphalt, amber, pine resins, tar oils, numerous vegetable oils and animal fats, soap, starch and gums. Vinegar and other organic acids were obtained from fruits, and tannins from bark. Dyes from madder, indigo, saffron and other plants were widely used.

The progress of chemical techniques led in the Middle Ages to the recognition of what were believed to be more "elementary" chemical substances—substances we now call sulfuric and nitric acids, saltpeter, ammonium carbonate, lead and mercury nitrates, and several acetates. "Combustible spirits of wine" (ethyl alcohol) was also recognized. The introduction of good distilling apparatus made it possible to prepare ethereal oils from plants—rose, violet, lavender, mint, aniseed, clove and many more. The flourishing metallurgical techniques of the sixteenth century added further inorganic substances such as the chlorides, nitrates and sulfates of gold, silver, iron, copper, tin, lead, mercury and antimony. Organic acids were isolated, such as succinic and benzoic. In the seventeenth century, Rudolf Glauber made many additions to the knowledge of chemical substances—ammonium nitrate, potassium chlorate, permanganate, benzene, ethyl chloride, acetone, strychnine and other compounds.

The substances listed above represent only a small sample of the multitude of materials that the chemist had to recognize, to describe, to name and to classify. As with plants and animals, there was at first no system in his procedures. But as techniques developed, chemists learned which properties were significant in physically manipulating substances—and hence of value in conceptually classifying them. In the tenth century, al-Razi used the properties of volatility, fusibility and solubility to distinguish between the volatile "spirits" (that included mercury and sulfur), the fusible metals, the soluble vitriols (such as alum) and salts, and the involatile, infusible and insoluble "stones" (including pyrites, mica, glass). Georg Agricola in *De natura fossilium* (1546) used the characteristics of fusibility, solubility, color, odor and taste to classify minerals.

One of the first European chemists to attempt a classification of all the known types of chemical substance was the German Johann Becher in *Epistolae chymicae* (1673). He recognized "minerae" (ores); metals (gold, silver, tin, copper, lead, iron); "mineralia" (including bismuth, zinc, antimony, magnesia); salts; "decomposita" (many inorganic materials); earths (mostly oxides); "distillata" (mineral acids, ammonia, spirits of wine); oils; "limi" (various minerals); and "compositiones" (secret preparations). In 1708, G. Roth in *De salibus metallicis* (1708) distinguished between "saline" substances (acids and alkalis), salts, and combustible substances—volatile (oils and spirits), less volatile (expressed oils, resins) and earthy.

Although chemists by the early eighteenth century had begun to put substances into some sort of classified order, they had not yet created a systematic nomenclature whereby the position of a substance in that order was revealed by its name. Indeed names—often based on superficial properties—sometimes led to mistaken groupings. For example, the "butters" (chlorides) of antimony and arsenic were classed along with ordinary butter; "oil of vitriol" (sulfuric acid) with olive oil; and "spirit of salt" (hydrochloric acid) with "spirit of wine" (alcohol). Substances were often named after their discoverers—Glauber's salt, alkahest of Van Helmont, Lilium Paracelsi, Boyle's fuming liquor, pearly matter of Kerkringius. There was even a "philosophical wood" (sublimed zinc oxide).

As well as "natural" names, the Renaissance alchemists developed a large number of symbols to represent not only substances but also operations and apparatus. Such symbolism had a long history. From Babylonian times, seven symbols had been associated with both the planets and the metals. In medieval and Renaissance times, the number of symbols greatly increased. The German chemist Oswald Croll's *Basilica chymica* (1609) gave signs for vinegar, salt, arsenic, sulfur, water, antimony, spirit of wine, borax, soap, sal ammoniac, saltpeter, alkali and so on. In France, Nicolas Lémery in *Cours de chymie* (1675) listed characters for substances and also for operations such as calcination, coagulation, distillation and filtration, and for apparatus such as alembic, crucible and retort. The symbols used were arbitrary combinations of lines, curves, circles, diamonds, crosses, crescents, triangles and letters. Johann Sommerhoff in 1701 published a chemical lexicon that included fifteen pages of chemical names and corresponding symbols. Some later chemists tried to introduce some system into the symbolism—e.g. the Frenchman Etienne Geoffroy in 1718 represented all mineral acids by various modifications of a circle; the Swede Torbern Bergman in 1775 used a distinctive sign to represent "calx," just as we use the letter *O* to represent its modern equivalent "oxide," and he combined this with the signs for metals.

THE CHEMICAL REVOLUTION

The new ideas introduced into chemistry by Antoine Lavoisier at the end of the eighteenth century found expression in a sweeping reform of nomenclature. To earlier chemists, when a metal was calcined something was emitted—"phlogiston"—so that a metal was a compound of phlogiston and its calx (despite Bergman's symbolism suggesting that the calx was compound). Lavoisier recognized that the calx was heavier than the calcined metal, and must be a compound of the metal and another substance. He identified this other substance as the gas discovered in 1774 by Joseph Priestley, and named it "oxygen." The new explanation had a profound effect on the interpretation of all chemical changes in which oxygen or oxygenous compounds took part.

Lavoisier himself popularized the new ideas in his *Traité élémentaire* (1789), but the basic book on nomenclature was published two years earlier, written by Guyton de Morveau with the collaboration of Lavoisier, Berthollet and de Fourcroy, the *Méthode de nomenclature chimique*. To it was appended *Un nouveau système de caractères chimiques*, by J.H. Hassenfratz and P.A. Adet.

De Morveau and his collaborators laid down principles of chemical nomenclature: a name should indicate a substance, define it, recall its constituent parts, classify it according to its composition, and indicate the relative proportions of the constituents. When nothing was known of the constitution of a substance, a name expressing nothing was preferable to one that might express a false idea. Attention was focussed on the simple substances, such as chemists had not been able to decompose, and about fifty-five of these were recognized—some now known to be chemical elements (e.g., hydrogen, oxygen, nitrogen, carbon, sulfur, the metals), others that were then considered simple "in the present state of knowledge" (such as lime, soda, potash). Rules were established to systematically build up compound names from the simple, thus "white arsenic" became oxide of arsenic; "martial aethiops" became oxide of iron; "vitriolic acid" became sulfuric acid; "Glauber's salt" became sulfate of soda. The old "natural" names were swept away in favor of names that expressed precisely the constitution of substances.

The symbolism proposed by Hassenfratz and Adet represented the simple substances by simple signs—e.g. oxygen as a horizontal line, nitrogen as a slanted line; hydrogen, carbon, sulfur and phosphorus as semi-circles in four positions; metals as circles with inscribed initials. Compounds were represented by combining these signs. The system was systematic, but clumsy, and did not indicate the relative proportions of constituents. In 1808, the English chemist John Dalton introduced the theory that atoms combined with each other in simple numerical proportions—A+B, A+2B, A+3B, etc.—and devised a symbolism to represent this: for each atom he used a circle in which was inscribed either a sign (e.g., a line for nitrogen) or a letter (e.g., C for copper), and a compound like nitric oxide was represented by a nitrogen sign linked to two oxygen signs. The last step towards modern chemical symbolism was taken by the Swedish chemist J.J. Berzelius in 1811, who replaced Dalton's signs by the initial letter (or letters) of the Latin name of each element, and indicated the number of atoms by a little figure placed to the right of the letter. He used a superscript "little figure," later to be replaced by the current subscript.

EARLY SCIENCE IN NORTH AMERICA

Let us turn now to some consideration of the spread of science. The first region overseas in which European science found a foothold was North America. In an earlier section, we have noted the Spanish, English, French and Dutch settlements during the period 1565–1625. The English came in the greatest numbers and—except in Quebec and Mexico—had the greatest influence on the process of development. Agriculture to supply subsistence was naturally the first concern. In the southern states, tobacco became a commercial crop almost immediately, later followed by rice and cotton, which (particularly after 1700) would be supported by the labor of African slaves. By 1700, the population of the English colonies was about a quarter of a million. During the eighteenth century, they pushed west to the Mississippi, and won independence from Britain.

Even in the seventeenth century, some settlers showed interest in science: for example John Winthrop, governor of Connecticut, was one of the founding members of the Royal Society. At first their main interest was in natural history, no doubt because of the obvious need to explore the natural resources of the country, but also because lack of equipment and lack of contact with European scientists made it difficult to advance physical science. The first learned society in the colonies was probably the Boston Philosophical Society, founded in 1683, led by the scholar Cotton Mather. This city was also the first in America to provide a public library, in 1653.

During the eighteenth century, contact with Europe was aided by such men as Peter Collinson, a London merchant and botanist with extensive trade connections in American towns. He acted as an intelligencer, linking American and European scientists, responsible for the publication of many American papers in the *Philosophical Transactions*, and shipping apparatus to America. One of his contacts, the statesman and scientist Benjamin Franklin, founded in Philadelphia in 1727 a small private library and scientific society, referred to as the Junto. Later, Franklin put forward a proposal "for propagating useful knowledge among the British plantations in America," and this in 1743–44 led to the formation of the American Philosophical Society, which was interested in botany, medicine, geology, geography, agriculture, mathematics, chemistry, and the crafts. Regular transactions were not published until 1771. The first major effort of scientific collaboration in America in which the society took part was the observation of the transit of Venus in 1769. An observatory was built in Philadelphia, and from there and other points in America the observations were made.

During the war for independence from England, French influence increased, and the idea of an academy of sciences on the French model came to the fore. In Boston, John Adams (later second president of the United States) formed the American Academy of Arts and Sciences in 1780. George Washington and Benjamin Franklin were among its first fellows, and it had a number of distinguished European members. Towards the end of the century, other societies were founded, such as the oldest state academy, the

Connecticut Academy of Arts and Sciences (1799). The first specialist societies were medical. Americans during the eighteenth century were going in increasing numbers to Edinburgh, Leyden and other European medical centers for training. Returning to their native land to practice, they set up societies—Boston had one as early as 1735. An Anatomical Society was founded at Harvard in 1771, and the Harvard medical school opened in 1782. Massachusetts had a medical society by 1781.

The expansion of manufacture during the latter half of the century led to the foundation of the New York Society for the Encouragement of Arts, Manufacture and Commerce in 1754. Agricultural societies were formed in Philadelphia (1785), Massachusetts (1792) and elsewhere. The credit for the first chemical society in Philadelphia is given to James Woodhouse (1792), and the American Mineralogical Society was founded in New York in 1798.

Section 6

The Nineteenth Century

THE DEVELOPMENT OF INDUSTRY

The nineteenth century saw modern industrial development really come into its own. Germany and France began to challenge Britain's initial industrial leadership, as shown in table 9.

In other European countries such as Sweden, in Russia, in the United States, and by the end of the century, in Japan, industry grew in importance. In the new economy, machinery began decisively to replace handicraft, and the machines were driven by power—not only by wind and water, but by coal-based steam power and, later, by electricity. In 1859, the first oilwell was drilled in Pennsylvania, and in ensuing decades wells were opened all over the world, e.g., Galicia, Romania, Russia (Baku), Java, Borneo, Burma. Manufacturing establishments grew in size—the factory became the typical industrial unit, and the large joint-stock company the typical business unit. The names of many of the entrepreneurs, engineers, and scientists who became industrialists are still associated with the products of today, for examples see table 10.

Industry began to provide a new occupational role and environment for scientists. Industrial research and development laboratories emerged—an early example being those of the prolific inventor Thomas Edison at Menlo Park, New Jersey, in 1876. In Germany, as early as 1868, German dye manufacturers were employing chemists, and by 1900 there were some four thousand trained chemists in the German chemical industry. The firm of Bayer by 1881 was employing fifteen chemists, and this increased to one hundred by 1896, together with a staff of technicians and a library. By 1884,

Table 9 Total Industrial Production (Britain in 1900 set at 100)

	1750	*1800*	*1830*	*1860*	*1880*	*1900*
Britain	2.4	6.2	17.5	54.0	73.3	100.0
France	5.0	6.2	9.5	17.9	25.1	36.8
Germany	3.7	5.2	6.5	11.1	27.4	71.2

Source: Bairoch (1982)

Siemens-Halske had a small physics laboratory, and at the same date, it was claimed, in England there were scarcely any metallurgical works without an analytical laboratory. By 1890 the English firm Read Holliday had a chemical research laboratory. Burroughs Welcome (1894) and British Westinghouse (1899) set up laboratory research. In the United States, from the 1870s, firms such as the Pennsylvania Railway Company, Carnegie Steel and General Electric were developing sustained research-and-development

Table 10 Entrepreneurs, Engineers, Scientists

Charles Tennant, bleaching powder (1825)
Samuel Morse, telegraph (1843)
Charles Goodyear, vulcanized rubber (1844)
Alfred Krupp, steel (1847)
Henry Bessemer, steel (1856)
Ernest Solvay, soda (1863)
Alfred Nobel, dynamite (1867)
P.D. Armour, meat canning (1868)
Ludwig Mond, alkalis (1873)
Philo Remington, typewriter (1873)
Alexander Bell, telephone (1876)
Ernest Siemens, electric generator (1876)
Johann Baeyer, dyestuffs (1880)
Charles Dunlop, pneumatic tire (1883)
Gottlieb Daimler, automoblie (1885)
Charles Parsons, steam turbine (1888)
George Eastman, camera (1888)
Sebastian Ferranti, electrical devices (1889)
Rudolph Diesel, petrol engine (1893)
George Westinghouse, generator (1893)
Auguste Lumiere, cine camera (1895)
Guglielmo Marconi, wireless (1901)
Henry Ford, automobile (1903)

programs. In 1872 Andrew Carnegie admitted the first professional chemist ("a learned German") to his Pittsburgh steelworks.

More evidence of the penetration of science into industry comes from the formation of industrial libraries, as in the German firm of Bayer. In Britain, libraries were formed in the chemical firms of Levinstein (1870), United Alkali (1891), Albright and Wilson (1899) and Nobel (1910). In the United States, some industrial library foundation dates are: Mallinckrodt (1867), Lawall (1870), Allied Chemicals (1879), A.D. Little (1886), Abbott (1888), Lilly (1891), General Electric (1900), Heinz (1903), Du Pont (1907), Babcock and Wilcox (1907) and Bell (1908).

Industry also showed its interest in science by its support for scientific establishments. For example, the chemical laboratory of Britain's Royal Institution was endowed by the industrialist Ludwig Mond, and by the end of the century up to two hundred thousand pounds per annum was being fed into the British university system from industrial firms. Such support was even more marked in the United States, where for example the University of Chicago was by that time receiving over a million dollars a year from industry.

One further feature of the nineteenth century will be mentioned: the development of transport and communications. At the beginning of the century, Richard Trevithick was experimenting with a steam locomotive that ran on rails in Wales, but it was in 1814 that George Stephenson built his first locomotive. This was followed up by a second in 1825 for a line between Stockton and Darlington, and by a third, the "Rocket," in 1830 to link Manchester and Liverpool. Railway construction in England expanded, with lines covering 6,000 miles by 1850 and 16,000 by 1874. This stimulated a great increase in passenger transport and linked the different regions of the country into an integrated economy. It also increased the demand for coal and iron and led to the further development of engineering.

In Europe, Belgium was one of the first to foresee the benefits of rail travel and by 1844 had a considerable system in place. France, Germany and later Russia followed suit. Each country was brought into a common economic framework. By the 1850s railway construction had become a major industrial activity in Europe. Between 1840 and 1880, some 80,000 miles of rail were laid in the continent.

The same development occurred in America: from 3,000 miles of rail in 1840, the network grew to 30,000 miles in 1860, and 166,000 miles in 1890. The railway was also brought to other countries outside Europe. For example, the first line was constructed in India in 1853, and the Indian mileage reached 25,000 by the end of the century. Between 1840 and 1880, world railways expanded to 230,000 miles, served by over sixty thousand locomotives. At one time the British firm of Thomas Brassey employed eighty thousand workers on railway projects over five continents. At the same time, the use of steam for propulsion and iron or steel for construction greatly enhanced sea transport.

Through rail and the steamship, European industry gained access to vast new land areas and markets. There followed the rapid colonization and cultivation of the United States west of the Mississippi; of central and western Canada; of the frontier country of Argentina, Uruguay, and Brazil; of the hinterland of South Africa and Australia—leading to copious supplies of cheap wheat, cotton, wool, mutton, beef, and leather.

The railways were the first users of a new development in communications—the telegraph. Early experimental forms of this were being tried out in the eighteenth century, but it was William Cooke and Charles Wheatstone who produced a practical instrument that they demonstrated in 1837 to railway managers. The code used to transmit messages was that developed by the American Samuel Morse. Telegraph-line construction proceeded not only in Britain, but also on the continent of Europe and in the United States. A submarine cable across the English channel was laid in 1851, and—after various attempts—an Atlantic link was also established in 1866. By then the world's telegraph system covered over 150,000 miles, and soon all the principal cities of the world were linked.

The transmission of speech came next—in the form of the telephone invented by Alexander Graham Bell in America (1876). The first telephone exchange was opened in London three years later. The German scientist Heinrich Hertz studied electromagnetic waves during the 1880s, but it was not till 1896 that Guglielmo Marconi, an Italian working in England, demonstrated his wireless apparatus to the Post Office. In 1901 he transmitted a wireless signal across the Atlantic.

The century also saw the industrialization of a traditional communications industry—printing. Improvements in the printing press were made by Friedrich König (1814). The first effective machine to cast metal type was developed by David Bruce in America in 1838. There were various attempts at producing a composing machine during the century, but success was eventually achieved with the production of the Linotype (Ottmar Mergenthaler, 1888) and the Monotype (Tolbert Lanson, 1889). Meanwhile, experiments on various forms of photography had been going on—for example, those of Joseph Niepce, L.J.M. Daguerre and Fox Talbot in the 1830s—and this technology was eventually to be used in photolithography, photogravure and the halftone process for producing illustrations.

TRANSFER OF TECHNOLOGY

The transfer of technical skills from one geographical location to another has gone on throughout history. For example, all the following technologies are believed to have reached Europe by diffusion from China: the horse stirrup and collar harness, the stern-post rudder, gunpowder, the mechanical clock, the navigational compass. The transfer agents were merchants and other travelers. Until the last few centuries, most technical knowledge was not recorded, and many artisans were not literate. Consequently, the transfer mechanism had to be either observation of a technique by a traveler who passed it on when he returned home, or—more effectively—by the movement of a skilled technician from one place to another. Pacey notes that the mere transfer of a novel product—or even the rumor of an unfamiliar technique—can stimulate technology. The knowledge that people in India could spin fine cotton yarns, weave delicate fabrics, and dye them with bright and fast colors stimulated British inventors to devise new ways of achieving the same results.

Technology is not simply a collection of equipment and materials, it involves also the experience and skills of those who operate the machines and work on the materials. The role of key personnel has often been a vital factor in successful technology transfer. This continued to be true in the eighteenth and nineteenth centuries, as a number of examples demonstrate.

In the 1750s, the English textile manufacturer John Holker moved to Rouen and set up weaving and spinning operations; a quarter of his workers were skilled British workers. A number of French employees were trained in Rouen and later became foremen at other new establishments. Several of the British workmen set up as builders of textile machinery in France. Holker went on to develop other areas of manufacture, again using British technology and British skilled workers. Another example is Samuel Slater, who moved from England to America in 1790, carrying in his head details of the spinning technology he had learned in England. Early in the next century, the inventor John Kay moved his spinning business to France, there first using machinery smuggled out of England by John Holker. In the 1820s, improvements in flax-spinning by Philippe de Gerard in France were transferred via personal networks to the Leeds industrialist John Marshall.

During the 1790s, the Scottish engineer Charles Baird was invited to Russia to establish an iron foundry and machine works. By 1825 he had constructed 140 steam engines, and was receiving many contracts. During the 1800s, the German engineer William Müller traveled throughout France,

Holland and Austria, inspecting advanced technology wherever he went; he then settled in Britain, and took out a series of patents on what he had learned. The graduates of French technical academies poured out into the railway systems of the Mediterranean countries, Russia, Austria, Switzerland and elsewhere. In Italy, French engineers drained harbors, built ports, canals and bridges. In 1843, William Siemens arrived in England from Germany with a new method of gilding by electrodeposition, which was bought from him by a Birmingham firm. He stayed on to develop the regenerative furnace.

As regards the world as a whole, one estimate of emigration from seventeen European states between 1846 and 1932 gives a figure of 51 million (18 million from Britain, 10 million from Italy, 5 million from Germany); the main recipients of this flow were the United States, Argentina, Canada, Brazil, Australia, South Africa and the West Indies. Emigrants carried with them a general knowledge of the manufacturing system and specific technical skills. Personal information transfer was an important factor in the communication of technical knowledge.

Important in a different way in the development and transfer of modern technologies has been the system of patent law. An early system was developed in Venice. In England, the first patent dates from 1552, but a reformed system was set up in 1624, extended to Scotland in 1707. An American patent law was passed in 1790, followed by France in 1791. In Germany there was no general validity of patents till 1842. In 1883 an international convention was agreed, covering the granting of patents in several countries. The number of patents granted in three leading industrial countries from 1842 to 1894 is shown in table 11.

In Britain in the 1850s, a series of government actions were taken to make the information in patents more widely available. The particulars of every important patent from 1617 on were reprinted, and copies of the specifications were placed in some 150 free or official libraries throughout the country; and a library was established in the Patent Office itself, to hold

Table 11 Patents Granted

Years	*Britain*	*Germany*	*USA*
1842–60	22,300	5,000	31,200
1877–82	22,100	21,100	50,000
1883–88	43,000	25,200	82,800
1889–94	59,300	32,500	77,500

Source: Inkster (1991)

not only the patents, but also "the scientific and mechanical works of all nations." These collections became much used by practical artisans. Between 1854 and 1857, the Salford library, which held twenty-three thousand specifications, reported 12,000 uses; between 1856 and 1859, the Manchester library reported 115,000 uses. Journals sprang up to report news of new patents: the *Artisan* provided full engravings and descriptions of new inventions and critical accounts of new patents; the *Mechanical Gazette* published a register of new patents and of machinery available for sale; other titles were the *Mechanics' Magazine*, the *Practical Mechanic's Journal*, the *Repertory of Patent Inventors*, and the *Patent Journal and Inventors' Magazine*. In the 1850s perhaps half a million British people were paying members of one or another institution for the diffusion of knowledge.

No other European country provided such a source of information—even the French Société d'Encouragement pour l'Industrie Nationale had a much more limited effect.

THE ORGANIZATION OF SCIENCE

The patronage of dukes, princes and monarchs was important for science throughout the sixteenth and seventeenth centuries, and even beyond. It was only at the very end of the eighteenth century that formal state support for institutions relevant to science began to grow.

In France, an École des Ponts et Chaussées had existed since 1747 to train civil engineers. In 1793, Lavoisier put before the French National Convention a comprehensive scheme for education in the arts and sciences, and the following year the École Polytechnique was opened, covering engineering, mechanics and chemistry. Here for the first time engineers were trained in fundamental science using laboratories and experiments. Its *Journal polytechnique* became a publication of world importance. At the same time, the Conservatoire des Arts et Métiers was set up as a museum of applied science, technical library and research station, and the old royal gardens were reorganized as the Muséum National d'Histoire Naturelle.

German universities throughout the nineteenth century were mainly state-funded, and much more professionally oriented than those of other European countries. To these during the 1820s were added technical high schools throughout the country. In these academic establishments well-equipped chemical and, later, physical laboratories were set up—one of the earliest being that of Justus von Liebig at Giessen. Lorenz Oken was instrumental in 1822 in founding an association of German-speaking scientists and doctors, the Deutscher Naturforscher Versammlung. This organization introduced the practice of national scientific congresses, which was taken up later in many other countries.

In Britain in 1799, the Royal Institution was founded. This was not in fact a state institution, but owed its origin to the enthusiasm of an American who had worked for the Bavarian state—Benjamin Rumford. Lecture and exhibition rooms, workshops and laboratories were provided. The Royal Society had been going through a long period of inaction. In 1820, the chemist Humphrey Davy—by then director of the Royal Institution—was elected president of the Royal Society, but the society was slow to take up its duties once more as a leader of British science. In 1828, Charles Babbage (later to develop the automatic calculator) attended a congress of the Deutscher Naturforscher Versammlung, and returned to England to write a book, *Reflections on the Decline of Science in England.* Scientific research, he claimed, was still largely amateur, neither state-aided nor professionalism. He and the physicist David Brewster founded the British Association for the Advancement of Science in 1831.

From 1823, mechanics institutes were formed throughout Britain, beginning with the London institute formed by George Birkbeck, and eventually numbering seven hundred. Their original intention was to give the elements of scientific knowledge to artisans and mechanics through classes, lectures and libraries (by 1850, the 610 English institute libraries owned seven hundred thousand volumes and circulated nearly 2 million items per year). But the original membership was eventually swamped by business and professional men and their families.

From 1826, the colleges of the University of London began to develop scientific facilities along the German university model. Government support for science in Britain was at first piecemeal. There had been a Board of Longitude in existence since 1714, and the state had financed expeditions that had employed geologists, oceanographers and astronomers. Scientists were employed at Kew Gardens, Greenwich Observatory and the Assay Office, the Museum of Economic Geology and the Geological Survey. Magnetic observatories and botanical stations were established throughout the British empire, and many geological surveys were undertaken in the colonies.

In the United States, the first center of training for engineers was the military academy at West Point (1794), followed by the Rensselaer Polytechnic in New York (1824). Massachusetts Institute of Technology was established in 1861. This acted as a model for a series of so-called "land-grant" colleges for agriculture and the mechanic arts, created during the 1860s. Meanwhile, other important scientific institutions had been formed, for example: the U.S. Patent Office in 1836, the Smithsonian Institution in 1846 (later including the Museum of Natural History and the Museum of the History of Technology), the American Association for the Advancement of Science in 1847, the U.S. Department of Agriculture in 1862, and the National Academy of Sciences in 1863. Libraries were established that would later be of great importance in scientific and technical communication: the Library of Congress (1800), that of the Department of Agriculture (1839), and that of the Army Surgeon General's Office (1836), later to become the National Library of Medicine.

The century witnessed a steady growth in all forms of scientific organization, and ever-increasing specialization. The number of societies and other institutions relevant to science, and of journals, becomes too great to follow in detail. The process will be exemplified by looking at the situation in Britain. Appendix A lists a sampling of nineteenth-century developments there. Similar lists could be compiled for a number of European countries and for the United States. The list illustrates a continuing growth in the formation of engineering and other technical institutions and journals, specialization of biological and medical societies, and the formation of new university and technical colleges, of medical schools and of museums.

NEW DIRECTIONS IN BIBLIOGRAPHY

Bibliography in the nineteenth century faced new problems. The volume of publication was tremendously increased. Iwinski estimated that nearly 9 million book titles had been published by 1900 (in all subjects), two-thirds of them in the nineteenth century (table 12).

An even greater impact on the bibliography of science was made by the development of journals. Some previous attempts had been made to record journal articles, noted earlier, but the problem grew rapidly more formidable. About two million scientific and technical papers were published during the century. It is clear that the age of the great individual

Table 12 Book Publication (in Thousands)

Century	15th	16th	17th	18th	19th
Books	30	242	972	1,637	6,100

Source: Iwinski (1911)

bibliographer was drawing to an end. A notable work of the early nineteenth century was the *Repertorium commentationum a societatibus litterariis editarum*, by the librarian of the University of Göttingen, Jeremias Reuss (1801–21). This was a subject index of the articles in the publications of academic societies up to the year 1800, fourteen of its sixteen volumes dealing with science, technology and medicine, and covering seventy thousand papers.

A project was suggested in the 1850s (at first by the American scientist Joseph Henry): to produce a catalogue of scientific papers that had appeared since 1800. At first adopted by the British Association for the Advancement of Science, the work was eventually undertaken by the Royal Society. As a result, the *Catalogue of Scientific Papers* was eventually published during the period 1867 to 1925, an author index to over fifteen hundred periodicals covering the whole nineteenth century, with about three-quarters of a million papers cited. It had been agreed from the beginning that a subject index should be provided, but only the subjects of pure mathematics, mechanics and physics were covered, less than a sixth of those in the author catalogue. To complete the story, the Royal Society published an annual *International Catalogue of Scientific Literature*, covering the years 1901 to 1913, but this became a casualty of World War I.

An enduring scientific bibliographical work of the mid-century is the *Biographisch-literarisches Handwörterbuch zur Geschichte der exacten Wissenschaften* of the German physicist Johann Poggendorff, the first volumes of which appeared in 1858. The work listed alphabetically the names of many thousands of mathematicians, astronomers, physicists, chemists, mineralogists and geologists, "of all peoples and times," each with a few lines of biography and a list of published books and articles. The main work covered scientists up to 1857, but since then other workers have produced supplementary volumes for later periods.

Two late-nineteenth-century specialist bibliographies may be mentioned. The Belgians J.C. Houzeau and A. Lancaster produced a multivolume *Bibliographie générale de l'astronomie* (1887), a systematic survey with a long historical introduction. A little later, in Washington H.C. Bolton published a *Select Bibliography of Chemistry 1492–1892*, covering twelve thousand items (1893).

However, the main bibliographical achievement of the nineteenth century was the development of the serially published record of scientific progress. As has been noted earlier, even in the eighteenth century there had appeared journals containing abstracts of recently published papers, for example the *Allgemeines Magazin der Natur, Kunst und Wissenschaften*

(1753 on), and later the *Esprit des journaux Français et étrangers* (1772 on). The *Berlinisches Jahrbuch der Pharmacie* (1795–1840) presented an annual narrative account of advances in its subject, though very few references were given in its early volumes. It was the forerunner of the modern annual review of progress. Almost contemporaneous was the *Bibliothèque britannique, ou recueil extrait des ouvrages Anglais periodiques et autres*, published in Geneva from 1796 to 1835, which offered annotated translations and popularizations of original scientific and technical papers. A further step towards the modern abstracts journal was taken in 1806, when the first volume appeared of the *Retrospect of Philosophical, Mechanical, Chemical and Agricultural Discoveries*, a quarterly published in London. The first issue abstracted seventeen patents and seventy-five papers from eleven journals (eight English and three French). At the end of the year, subject and author indexes were supplied.

A decisive advance occurred in 1822 when there was published the first volume of the *Jahresbericht über die Fortschritte der physischen Wissenschaften,* compiled by the chemist Jacob Berzelius and translated from the Swedish by C.G. Gmelin. In this, Berzelius reported on the year's achievements in physics, chemistry, mineralogy and geology, with about 170 references in footnotes. The *Jahresbericht* was one of a series of narrative annual reviews inaugurated by the Swedish Academy of Sciences, the others covering astronomy, botany and zoology. When Berzelius was writing his annual review it was his custom to stay away from the laboratory, surround himself with the literature he wished to abstract, and for several weeks almost withdraw from human society. This procedure was repeated annually for more than a quarter of a century. As his colleague Wilhelm Ostwald said, "it was under his diligent, faithful hands that the idea [of an annual report] became vitalised and grew to manhood, then made its way around the entire scientific world."

An early English attempt at an annual review was the *Arcana of Science and Art* (1828 on), which was "an annual register of the useful arts," providing précis of mainly British papers. In the 1840s, two new German annual reports appeared: *Die Fortschritte der Physik* was a classified survey of physics with a list of references, which continued as *Physikalische Berichte*; and Justus Liebig and Hermann Kopp founded the *Jahresbericht über die Fortschritte der reinen, pharmaceutischen und technischen Chemie, Physik, Mineralogie und Geologie*, the first volume containing two thousand references.

Despite these achievements, the increasing volume of scientific papers demanded surveys that appeared more frequently than once a year. In 1830

Table 13 Early Abstracting and Indexing Serials

American Journal of Pharmacy, 1829
Zentralblatt für Geologie und Paläontologie, 1830
Jahrbucher der gesamten Medizin, 1834
Retrospect of Practical Medicine and Surgery, 1840
Bulletin de la Société Chimique de France, 1858
Pharmazeutische Zentralhalle, 1859
Record of Zoological Literature, 1864
Jahrbuch über die Fortschritte der Mathematik, 1871
Index medicus, 1879
Botanisches Centralblatt, 1880
Biologisches Zentralblatt, 1881
Engineering Index, 1884
Anatomischer Anzeiger, 1886
Zentralblatt für Bakteriologie, 1887
Experiment Station Record, 1889
Science Abstracts, 1898

the first issue was published of the *Pharmaceutisches Centralblatt*, which continued as *Chemisches Zentralblatt*. A further form of publication was the *Archiv für Naturgeschichte* (1835 on), first edited by A. Wiegmann, a bimonthly consisting of review articles on aspects of biology. At least 150 abstracting and indexing serials were founded during the century, some with short lengths of life. Some of those that continued longer—even until today—are shown in table 13.

Section 7

The Twentieth Century

THE ENVIRONMENT OF TWENTIETH-CENTURY SCIENCE

The industrial revolution that began in the 1750s has transformed the world. In the countries that have achieved full industrialization, the production and consumption per head of goods and services is now at least ten times greater than in preindustrial societies. The balance between rural and urban living has been vastly changed—a far greater proportion of people now live in cities, and cities are larger. In 1750, the largest city was probably Canton or Constantinople (about a million people). Today it is Tokyo or Mexico City, with over 20 million, and there are many cities with over 5 million. Numberless new products have been created during the twentieth century—the portable radio, television, computers, artificial fibers, plastics, antibiotics, tranquilizers, and thousands more. Transport and telecommunications have been revolutionized. Following the geographical discoveries of the fifteenth and sixteenth centuries, European explorers, traders, soldiers and investors spread throughout the world. The goods, technologies and ideas of the industrial West came to be found everywhere. World trade greatly expanded, as the data in table 14 show.

One pervasive feature of the present era has been that of uneven development. This applies both to social conditions between countries (which will be considered in the next section) and to those within each country. The main institution to display uneven development has been the business firm. From a situation in which few firms were of any consequential size, unequal rates of growth, mergers and acquisitions during the late-nineteenth and twentieth centuries led to the creation of ever larger enterprises.

Table 14 Volume of World Trade (1750 set at 100)

Decade	*Volume*
1750s	100
1820s	206
1830s	268
1840s	433
1850s	692
1860s	1,056
1870s	1,623
1880s	2,286
1890s	2,873
1900s	4,083
1910s	5,378
1920s	6,077
1930s	6,131
1950s	7,045
1960s	13,122

Source: adapted from Landes (1969)

Big firms developed early in transport, particularly railways. In the early 1900s, large automobile manufacturers became prominent through the merging of smaller firms (Morris, Austin, General Motors, etc.). Large steel firms were formed, followed by chemicals—for example, in Britain in 1926 Imperial Chemical Industries resulted from the merger of United Alkali, Brunner Mond, British Dyestuffs, Nobel and Kynoch, and in 1929 Unilever was created. Soon it was the turn of the oil companies. To give a dramatic picture of the size of major corporations, it became common to compare their annual sales with the gross national products of various countries, as for example in figure 7, taken from Dicken and Lloyd. By 1990, there were a dozen firms with over two hundred thousand employees (including General Motors, Ford, General Electric, Daimler-Benz, Philips, Siemens, IBM, Unilever, British-American Tobacco, Hitachi, Fiat and Volkswagen).

Such corporations were soon no longer restricted to one country, but became multinational, with operations worldwide. Fourteen British firms were operating transnationally by 1913, including some in high-technology areas such as oil (Shell), synthetic textiles (Courtaulds), rubber (Dunlop) and armaments (Vickers). At the same date, seventeen European and over thirty American companies had overseas interests. MacBride listed fifteen

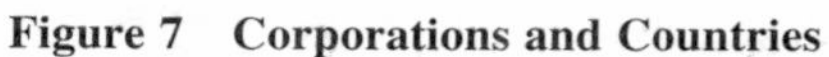

Figure 7 Corporations and Countries

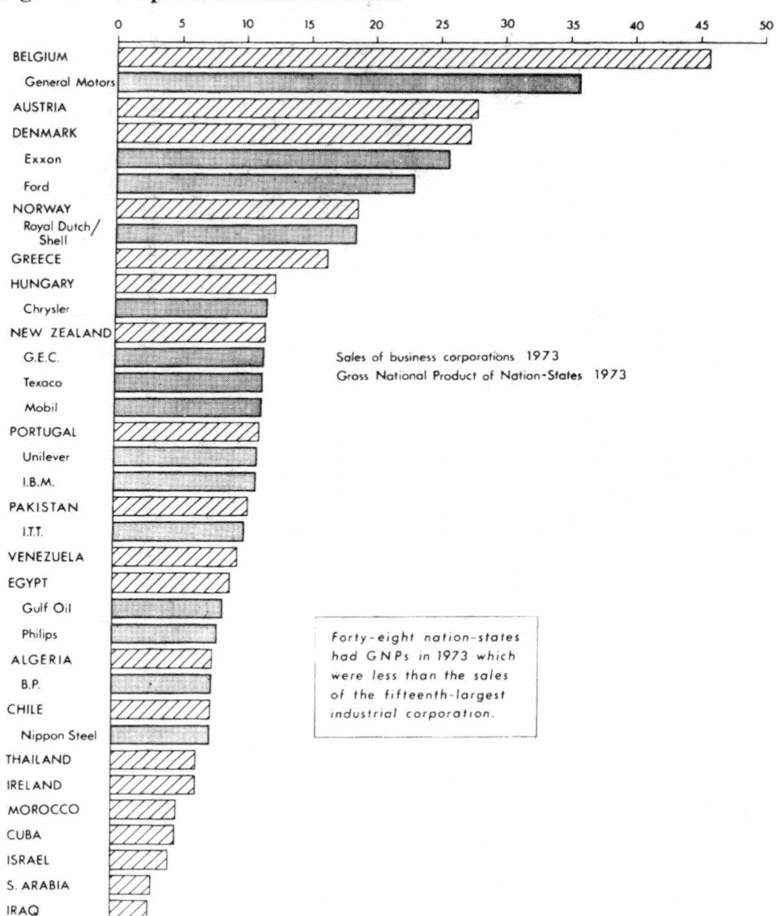

transnational corporations that in 1980 controlled in different ways the largest part of operations in international communication (table 15).

Large corporations came to provide an important new work environment for scientists engaged in industrial research and development, particularly but not only in the United States. In a previous section we have looked at late-nineteenth-century developments. In the United States, starting with General Electric in 1900, and followed by DuPont, American Telephone

Table 15 Transnational Corporations, 1980

Corporation	*Headquarters*	*Employees*
IBM	United States	289,000
General Electric	United States	375,000
ITT	United States	376,000
Philips	Netherlands	397,000
Siemens	Germany	296,000
Western Electric	United States	153,000
GTE	United States	187,000
Westinghouse	United States	166,000
AEG-Telefunken	Germany	162,000
Rockwell	United States	123,000
RCA	United States	113,000
Matsushita	Japan	83,000
LTV	United States	60,000
Xerox	United States	93,000
CGE	France	131,000

Source: MacBride (1980)

and Telegraph, Eastman Kodak and Westinghouse, big firms set up central research facilities. By 1927, some thousand industrial research laboratories existed in America. Large European and Japanese firms trod the same path.

Another major institutional environment for scientists that developed was the government agency, for example in astronomy, in aeronautics and space, in atomic physics, in nuclear energy, in medical research, in armaments. In Germany, the first national research institute was the Physikalische Technische Reichanstalt of 1887, formed by the state with financial support from the Siemens firm. In 1911 the Kaiser Wilhelm Gesellschaft was created, and today, as the Max Planck Gesellschaft, it has about sixty attached research institutes.

In the United Kingdom there was formed in 1916 the Department of Scientific and Industrial Research, with a number of associated research associations; later government institutions include the Royal Aeronautics Establishment and the U.K. Atomic Energy Authority; public corporations were formed during the century—in steel, coal, electricity, gas, railways—that were larger, in employment terms, than any private corporation; and a large national health service was created.

When the Soviet Union was formed, the reorganized Akademiya Nauk was made responsible for a very wide range of institutes. In France in 1939 was formed the Centre National de la Recherche Scientifique, again with a range of associated research laboratories. In the United States we may mention the National Institutes of Health and the National Aeronautics and Space Administration; it has been claimed that the U.S. Department of Defense's wholly owned manufacturing facilities at one time ranked fifth in employment terms among American industrial establishments.

A high proportion of the world's scientists and technologists (more particularly in the fully industrialized countries) came to be found in what Ziman called "the chunky institutions of Big Science." Scientists came typically to work in teams. Many no longer did individual research: they took part in research projects, collaborative efforts in which there was division of labor—even to the extent that a special group might be designated to collect and process published information for the rest of the team. Multiple authorship of research reports steadily increased during the century. Complicated and expensive equipment was often required to prosecute research. A corollary to these developments was that a relatively smaller proportion of the total research and development in an industrial country was performed in universities and colleges. For example, Bell reported that in 1965, of the half million U.S. research and development scientists and engineers, 13 percent were employed in universities and colleges, 14 percent by the federal government, and 70 percent in industry. In 1989, Freeman reported, there were 950,000 U.S. research-and-development scientists and engineers; the percentage in universities and colleges had improved to 18 percent, the federal government figure had dropped to 6 percent, and industry had risen to 76 percent.

There came about a complex interplay between basic scientific and technological development. This is suggested in general terms in our earlier figure 1, and in more detail in figure 8, showing the paths that led, from 1850 to 1960, to the invention of the videotape recorder (adapted from de Vore).

One last general feature of science may be mentioned—its continuing division into specialties. This came about in part as an inevitable response to growth. As the number of scientists in a subject area expanded, so did the content of scientific material that it covered, and it tended to break into two or more subdivisions—as, for example, chemistry became divided into physical, organic, inorganic, analytical and so on. A second mode of formation of specialties was a result of the methods of one science being applied to the material of another—so a chemical approach to living organisms gave rise to biochemistry. Third, new discoveries or inventions led to the creation of new specialties—for example, the discovery of colloids, or the invention of the computer.

Figure 8 Genealogy of an Invention

Themes Contributing to Invention of Videotape Recorder

I. Control Theory

1. Black–BTL–1920: feedback theory
2. Black, Nyquist, Bode–BTL–1932–34: control theory
3. MIT, BTL, NRL–1940s: servomechanisms
4. 1940s: precision mechanical engineering
5. 1940s: reliability and quality concepts
6. Brown, Campbell–MIT–1948: servomechanisms textbook
7. 1950s: control of rotating mechanisms
8. 1950s: lateral scanning with rotating heads

II. Magnetic and Recording Materials

1. Hadfield–UK–1882: hardness of iron-silicon
2. Poulsen–Denmark–1896: steel wire for recording
3. Barrett, Brown–UK–1900: improved magnetic materials
4. Mix, Genest–Germany–1900s: multitrack steel tape
5. 1910s: studies of magnetic materials
6. 1910s: studies of plastics for tape
7. Elmen–BTL–1913: iron-nickel permalloy
8. Pfleumer–Germany–1927: plastic tape for magnetic recording
9. O'Neill–USA–1927: two-layer paper tape
10. IG Farben–Germany–1928: improved plastic tape
11. 1930s: improved preparation of magnetic tape
12. late 1930s: improved binding and backing materials
13. AEG–Germany–1935: magnetophon tape recorder
14. late 1940s: two-layer tape perfected
15. 1940s: improved tape coating
16. Camras–IITRI–1946: improved magnetic properties
17. early 1950s: reliable wideband tapes
18. 1950–55: wideband video recording feasible

III. Magnetic Theory

1. Weber–Germany–1852: ferromagnetic theory
2. Ewing–UK–1885: hysteresis effect
3. Smith–USA–1888: recording on magnetically coated material
4. Weiss–ETH Zurich–1907: studies of magnetism
5. 1920s: much work on ferromagnetism

IV. Magnetic Recording

1. Poulsen–Denmark–1897: workable wire recorder
2. Mix, Genest–Germany–1910: recording on wire and steel tape
3. Stille–1920s: electronic amplifiers

4. 1930s: wire and steel tape widely used
5. Schuller–Germany–1930: ring magnetic head
6. AEG–Germany–1935: magnetophon
7. Lubeck–Germany–1937: magnetic recording theory
8. 1940s: improved signal-to-noise ratio
9. Camras–IITRI–1940: AC bias techniques
10. 1950s: tape recorders available
11. Wallace–BTL–1951: loss laws of magnetic recording

V. Electronics

1. Edison–US–1883: current flow in vacuum tube
2. Fleming–UK–1904: vacuum diode
3. De Forest–USA–1906: triode vacuum tube
4. Arnold–BTL–1905: improved triode
5. Schottky–1920: tetrode
6. 1920s: workable triode
7. 1934: pentode
8. 1930s: electronic industry evolves

VI. Frequency Modulation

1. Hertz–Germany–1887: verified Maxwell's equations
2. Marconi–UK, Poulsen–Denmark, Fessenden–USA–1900s: early wireless propagation
3. Ehret–1902: coded signals transmitted
4. Englund–Western Electric–1914: side bands
5. Goldsmith–CUNY–1918: FM telephony
6. 1920: FM becomes practical
7. Carson–USA–1922: band-width studies
8. Armstrong–192S: low noise characteristics of FM
9. Crosby–RCA–1936-41: phase and frequency modulation
10. Gambi–Italy–1948: mathematical analysis of FM
11. FCC–1949: TV transmission using FM

An illustration of the degree to which institutional specialization has proceeded is given in appendix E, which lists research institutes attached in 1990 to the French Centre National de la Recherche Scientifique.

UNEVEN DEVELOPMENT OF INDUSTRIALIZATION

The expansion of Western power throughout the world left no country untouched. But industrialization developed very unevenly. Many nations of the Third World have not yet had their industrial revolution, and the gulf in wealth and living standards between them and the industrially advanced countries has dramatically increased. It has been estimated that, whereas in the middle of the nineteenth century the living standards of western Europe were only 50 percent higher than in Asia, Africa and Latin America, by 1960 the gap had grown to eleven times higher. Lenski gave an example of some differences between a fully industrialized society (United States) and one less developed (India) in 1975 (table 16).

After 1960, in some respects the gap between fully industrialized and less developed countries began to decrease. This was particularly so in rapidly growing economies such as South Korea, Brazil, Taiwan, Mexico, Malaysia and the oil-rich OPEC countries. But overall, the disparity in gross national product per head (GNP) continued to increase. It should also be noted that the proportion of the world's population living in less developed countries is increasing: in 1850 the proportion was about one-half; demographers estimate that by 2050 it may be more like five-sixths.

There are cited in table 17 summary figures showing some recent changes for a number of sample countries (all figures are percentages relative to industrial countries, which are set at 100).

Relative to the industrialized world, real GNP actually decreased in countries such as Mexico, India and sub-Saharan Africa. In developing countries as a whole, the adult literacy rate increased by more than one-third over the period 1970–90, but there were still nearly a billion illiterate adults. The proportion of the population going to university increased during the same period—but one-third of adults had not completed secondary education. The average years of schooling in industrial countries in 1990 was ten,

Table 16 Comparative Standards in 1975

Index	*USA*	*India*
Infant mortality per 1,000 births	18	139
Hospital beds per 1,000 people	7.2	0.6
Doctors per 1,000 people	1.6	0.2
Energy used per head (1,000 kw)	12	0.2
Steel production per head (tons)	0.65	0.01
Cereal production per head (lbs)	2,024	352
Meat production per head (lbs)	194	31
Milk production per head (lbs)	550	88
Telephones per 1,000 people	657	3
Radios per 1,000 people	1,750	220
Televisions per 1,000 people	520	1

Source: Lenski (1978)

Table 17 Changes between 1960–70 and 1985–90

Country	*Period*	*Real GNP per head*	*Adult literacy*	*School enrollment*	*People per nurse*
Mexico	1960–70	46	56	77	48
	1985–90	38	74	86	16
Brazil	1960–70	21	69	75	15
	1985–90	33	83	91	12
Saudi Arabia	1960–70	77	9	36	8
	1985–90	69	64	65	41
India	1960–70	11	35	56	7
	1985–90	6	49	69	8
South Korea	1960–70	15	92	87	16
	1985–90	41	98	100	24
Egypt	1960–70	11	37	63	–
	1985–90	13	49	92	–
Sub-Saharan Africa	1960–70	14	28	30	12
	1985–90	8	53	47	6
Industrial countries	1960–70	100	100	100	100
	1985–90	100	100	100	100

Source: UN Human Development Report (1992)

in the developing world it had not reached four. The number of daily newspapers distributed per thousand population in Europe and North America was 250, in Latin America it was 70, in the Arab states 45, in sub-Saharan Africa about 13. The number of telephones per thousand people was nearly 800 in North America, 500 in the European Community, 85 in Latin America, 55 in the Arab states, and 15 in southeast Asia and sub-Saharan Africa.

The United Nations report indicates that the world's scientists and technicians in 1990 were distributed as shown in table 18. The third column shows the number of scientists and technicians per thousand people.

If we look now at research-and-development scientists, in industrialized countries there were about 5 million; in all the developing countries, about 1 million. Despite having 80 percent of the world's population, and apparently a sixth of the world's research-and-development scientists, developing countries were responsible for only 4 percent of global expenditure on research and development. This implies that the resources available per

Table 18 Distribution of Scientists and Technicians (Figures in Millions)

Area	*Scis and techs (A)*	*Populn (B)*	*1,000A/B*
Industrial countries	98	1,210	81
Latin America	13	440	29.6
Arab states	3.7	270	13.7
East Asia with China	12.1	1,210	10
South Asia with India	4.3	1,200	3.6
All developing countries	over 33	4,070	8
All countries (industrial and developing)	over 131	5,280	25

scientist in the industrial countries were nearly five times as large as those available per scientist in the developing countries.

Modern developments in informatics and communications seem to have bypassed many developing countries. Only a twentieth of the world's computers in 1990 were outside the industrial countries. The technological gap between industrial and developing countries widened in the period 1960–90, partly because scientific and technical information was becoming increasingly privatized. The application of intellectual property rights (embodied in patents) became more stringent; buying rights to technology was not the answer for many developing countries, since prices were beyond their means.

The United Nations and other agencies during the late twentieth century devoted much effort to "technology transfer"—the provision of technical expertise by industrialized countries to the less developed. The United Nations Development Program reported that such technical assistance programs had limited success in building up the national capacity of developing countries. Modern technology is often large-scale, complex and specialized, and less developed countries could not readily adapt to it.

Multinational corporations invested in some developing countries and set up manufacturing units in them. Insofar as such units employed local labor there was a degree of technology transfer through training in special skills and techniques. But the corporation might well restrict the spread of this technology to other firms in the country. The multinationals have located relatively few technology-creating facilities—that is, research and development—in developing countries, except low-level support laboratories.

The effect of these disparities was that the environment of science in the developing countries was considerably different from that in the industrialized world. Ziman suggested a typical picture of the situation in a poorly developed country: the modest institutions of higher education inherited from the past have been expanded into substantial universities crowded with students, though the curriculum might have barely changed for half a century, and each subject might be taught separately and dogmatically. The number of science students was relatively small, but for advancement to professorship there was a need to do "research." This could often only be achieved by undertaking advanced study in a university of an industrial country. So off might go the ambitious student, to spend four or five years working with sophisticated apparatus on a recondite specialist problem. Back home, he would like to continue research—but the laboratory resources were quite inadequate and teaching duties were heavy. There was relatively little effective government support for research—there was much talk of science and technology transforming the nation, but so far with little success. There might be one or more high-technology factories in the country, but these would often depend on foreign technology, with foreign experts in charge. Local scientists might rarely be employed to adapt existing techniques to local use, or to invent new products. As a result, the new Ph.D. might either succumb to his environment and concentrate on teaching and career advancement, or join the "brain drain." Developing countries lost many thousands of scientists, engineers, doctors and technicians each year: between 1960 and 1990, the United States and Canada accepted more than a million professional and technical immigrants from developing countries. Japan and Australia also tried to attract skilled immigrants.

This background must be remembered when we later discuss the modern forms of communication available to the scientist: these benefits were much less in evidence outside the industrialized countries.

SCIENCE OUTSIDE EUROPE

The science developed in Europe has nevertheless spread to a number of other countries, starting, as we have seen earlier, in North America. To give some indication of this process, developments will be briefly described in a few sample areas: Australia, Japan, Brazil, India and Africa.

Australia was the last overseas area to be discovered and settled by Europeans. Although the Dutch explorer Abel Tasman had discovered it in 1642, it was only the voyage around it of James Cook in 1770 that led to colonization by British settlers, displacing the aborigine population of hunter-gatherers. Australia at first lived on its agricultural and mining products, and manufacture was slow to develop. Sydney's botanic gardens opened in 1816. A Royal Society was founded in New South Wales in 1821, followed by such societies in the other states later in the century. The first universities were founded in the 1850s, and in time housed flourishing schools of science, engineering and medicine. Also during the 1850s the Geological Survey of Victoria was set up. The first specialist societies were those of the zoologists (1857), the field naturalists (1880), the astronomers (1892) and the mining engineers (1893). An Association for the Advancement of Science was formed in 1886. The state Departments of Agriculture acted as centers of research and employment for chemists, plant pathologists, entomologists and bacteriologists.

The early twentieth century saw the creation of societies of ornithology (1901), physiotherapy (1905), wildlife (1909), chemistry (1917) and engineering (1919). Other specialties followed. A Council of Scientific and Industrial Research (now CSIRO) was formed in 1926, and supported much research, particularly in agriculture. In the 1930s, a research institute of agricultural science and a soil conservation service were established. Manufacturing industry began to grow more quickly after 1945, but the development of industrial research was slow, though some industrial research associations were formed. By 1990, Australia had over fifty thousand research-and-development scientists, and fifty scientists and technicians per thousand of its people.

Japan offers a marked contrast, in that it was never colonized by Europeans. The earliest Japanese historical records are of the eighth century A.D., though reference to the country is found in third century Chinese texts. The Japanese writings used Chinese characters to represent their own very different language. In the sixteenth century Japan was dominated by land-owning lords, with central control established during the century by the Tokugawa family. In the towns, handicrafts and trade flourished. Portuguese explorers first made contact in 1543, and for a while began to exercise influence, but in the 1620s they and most other Europeans (except for a small Dutch settlement) were expelled—an exclusion that lasted until the 1850s, when the United States navy opened up the country to foreign traders. Soon after, the rule of the Tokugawa came to an end: power became centered in the emperor's court (as had been the case many centuries before)—the so-called Meiji restoration of 1868.

During the Tokugawa period, Japan had not been altogether isolated from Europe. In the eighteenth century, a number of Japanese scholars learned Dutch and studied European medicine, agriculture and other technical subjects. Translations were organized and widely read in a society with considerable literacy. Some young men were sent abroad to study Western technology. The interaction continued in the first half of the nineteenth century. For example, in 1854, a German book on new industrial technologies was translated, and a translation office was set up to put Dutch technical books into Japanese. In 1865, a state college was formed under the patronage of the American Rutgers University, "for the benefit of agriculture and mechanical arts."

The new Meiji rulers began rapidly to follow up these beginnings and to adopt European institutions and technology. Their charter proclaimed: "knowledge shall be sought throughout the world so as to strengthen the foundation of imperial life." The first railway, from Tokyo to Yokohama, was built in 1872 by foreign engineers directed by William Cargill. Universities were founded at Hokkaido in 1876, and at Tokyo in 1877, at first using as teachers such foreign experts as the engineer Alfred Ewing. Tokyo Institute of Technology was set up in 1881. Between 1868 and 1895, twenty-five hundred students and officials were sent to Europe and America to study technology. The Japan Academy was formed in 1879, and a number of societies were founded at this time: mathematics, medicine, zoology, chemistry, geology and engineering, followed in the 1880s by mining, agriculture, botany, meteorology and electrical engineering. Many industrial exhibitions were held.

After 1900, Japanese industry began to develop rapidly, as did its scientific activity. A survey in the 1950s identified over sixteen hundred scientific and technical institutions in Japan, over half in technology, most of the others in agriculture and medicine, only ninety in science. There were at the same time published nearly eight hundred journals in these subject areas. Less than half of the scientific journals used Japanese alone, the rest being published wholly or partly in English or other European languages; in technology, the proportion of wholly Japanese journals was higher. An Agency of Industrial Science and Technology had been set up in 1948, to which a number of research institutes were attached. It is generally held that scientific research in Japan up to that time, though copious, was not making major contributions to world science.

From the 1950s, the import of technology and the growth of science was even more rapid. Many hundreds of thousands of license agreements for Western technologies were signed, and during the 1960s perhaps a third of Japanese production was directly dependent on foreign technology. But by the 1980s Japan's export of technology began to predominate over its import. By 1990, Japan had nearly three-quarters of a million research-and-development scientists, and 110 scientists and technicians per thousand of its people.

Let us note now a Latin American country, Brazil. The country was first claimed by Europeans in 1500, when the Portuguese Pedro Cabral, intending to reach India via the south of Africa, sailed too far west in the Atlantic and encountered land. The native inhabitants had a primitive economy, not like the Aztec, Inca or Maya civilizations elsewhere in Latin America. Under Portuguese control for three hundred years, Brazil became an independent country in 1822. Like Australia, for a long time its economy was centered on agriculture and mining. An institute for agronomy was set up in 1887, and an agricultural society ten years later. An academy of sciences was established in 1916. Industry began to develop more strongly after 1940, together with specialist scientific societies: metals and chemical engineering in 1944, biology in 1947, chemistry in 1951. In the 1970s the Empresa Brasiliera de Pesquisa Agropecuária was set up, with a number of attached agricultural research institutes. During the 1980s, industrialization made rapid advances. In 1990, Brazil had perhaps seventy-five thousand research-and-development scientists and thirty scientists and technicians per thousand of its people.

The development of India has been very different from any of the examples so far considered. A civilization comparable to those of Mesopotamia and Egypt existed forty-five hundred years ago around the Indus river, though relatively little is known of it, and records of its written script have not been deciphered. Over the millennia, a rich and varied culture was created. From about 1500 B.C., the Indus valley was invaded by Aryan tribes, whose literary records are the four Vedas. In them, mathematical, astronomical and medical knowledge can be found. From about the seventh century B.C., formal philosophical systems began to be developed. A number system was created, that eventually—via the Arabs—gave rise to the modern numerals. Medical practice was codified into Ayurveda, "the science of life." In the fifth to seventh centuries A.D., astronomy and mathematics were further developed, and alchemy (which had been practiced since the Vedic period) was further extended. From the seventh century, contact with Arab traders led to the exchange of scientific and medical knowledge between Indians and Arabs. Indigenous Indian science did not develop greatly beyond the "medieval" stage it had reached in 1200. During the seventeenth century, there was some exposure to European science: for example, François Bernier, a physician from Montpellier, met Indian rulers and gave accounts of contemporary European work; in Jaipur, the local maharajah began to collect astronomical texts from India, Arabia and Europe, and set up five observatories.

The first British settlement in India was at Surat in 1613. During the seventeenth century, trade was developed, but little further penetration of the country took place. India was mainly agricultural, but there were flourishing craft industries in the villages and towns—weaving, pottery, cotton textiles, silks, metals, and other trades. These continued to develop during the century, under the rule of the Moguls. The standards of cotton materials and iron and steel work were reported to be higher than anything in Europe at that time. Royal factories were established, for the production of both weapons and commercial goods such as silk. By 1700, with the weakening of Mogul rule, Britain began a struggle both to displace commercial rivals—Portuguese and French—and to gain control of India. This process was completed by the end of the eighteenth century.

Under British control, the Indian textile industry, iron production and other technologies such as shipbuilding began to be eroded by cheap imports from Britain, and handicaps were placed on Indian merchants. When the construction of railroads started in 1853, Indian participation in any skilled role was excluded. By the end of the nineteenth century, India had a bigger length of railroad than the rest of Asia put together, but there was little related industrial and engineering development. Nevertheless, some independent Indian industry began to appear. For example, the Tata family founded a cotton spinning company in 1868, and later a steelworks: Tata Industries eventually became the largest industrial combine in India.

How did science fare during this period? In 1784, William Jones started the Asiatick Society of Bengal, and under his stimulus Sanskrit works on astronomy, and medicine were translated into English. The East India Company (the effective government of India) set up colleges and medical schools for Indians to study the Indian sciences in the vernaculars. A Royal Botanic Gardens was formed in 1787 at Sidpur, but this was regarded by the British mainly as a source of material for their own Kew Gardens. Support for indigenous Indian science was however short-lived: in 1833 all the native colleges were closed to make way for the teaching of science and medicine along Western lines, and in English—for example, at a new Calcutta Medical School. During the century, official survey organizations were set up—trigonometrical (1818), geological (1851), meteorological (1875), botanical (1890)—but Indians were excluded from this work. A natural-history society was set up in Bombay in 1883. Some independent Indian efforts took place to develop institutions to study Western science: in 1864 there was founded the Aligarh Scientific Society, and four years later another such society in Bihar; an Indian Association for the Cultivation of Science was set up in 1876. Towards the end of the century, the Bengal Association for the Advancement of Scientific, Agricultural, Industrial and Commercial Education began to send students abroad for further studies. But there was little encouragement of these efforts by the British.

Nevertheless, during the first half of the twentieth century, new scientific institutions were formed: the Indian Institute of Science (1909); a Council of Medical Research (1911); the Indian Science Congress (1914); a Council for Agricultural Research (1929); the Indian Statistical Institute (1931); the Indian Academy of Sciences (1934); and specialist societies in mathematics, chemistry, medicine, soil science, physics, engineering and other disciplines. The Council for Scientific and Industrial Research, with attached research institutes, was formed in 1942. India became independent of Britain in 1947. Since then, some outstanding individual scientific institutions have been formed—for example, the Tata Institute for Fundamental Research in Bombay—and Indian Institutes of Technology, founded in the 1960s to provide sophisticated engineering training. Little research in private Indian industry was undertaken. In 1990, India had nearly two hundred thousand research-and-development scientists (a large number, but low in relation to the population of 850 million), and 3.6 scientists and technicians per thousand of its people.

The term sub-Saharan Africa is used to refer to the whole of Africa south of the Sahara and encompasses all African countries except those on the Mediterranean (Egypt, Libya, Morocco, Tunisia and Algeria). The total population of the area is 500 million, and over forty different states are involved. One is reasonably industrialized—South Africa—but many are among those classed as the world's "least developed." Almost all have been colonies of Britain, France, Belgium, Portugal, Spain or Germany, and achieved independence in 1960 or later. An analysis of the *World of Learning*, 1991, for all sub-Saharan countries, excluding the industrialized South Africa, reveals about fifty scientific societies and twice or three times that number of research institutes. Taking 1960 as an average date for transition to independence, a rough pattern of start-up dates for societies and institutes is shown in table 19.

These figures indicate that little physical, chemical and technological research had yet developed in this huge area. The proportion of the world's foreign investment going into Africa declined during the 1980s, "a reflection," wrote Dicken, "of its increasing marginality in the global economy." Data for the number of scientists and technicians per thousand people are not available for the whole sub-Sahara: for Zambia the figure in 1990 was 4.4, for Botswana, Kenya and Ghana in the range 1.2 to 1.5, for Nigeria 0.9, for Rwanda 0.2, and no doubt less for some countries. The number of research-and-development scientists over the whole area was probably less than fifty thousand.

Table 19 Sub-Saharan Societies and Institutes

Subject area	*Before 1960*	*1960 on*
Science in general	10	10
Geology and geophysics	12	5
Meteorology	4	–
Agriculture, agronomy	20	20
Specific crops (cocoa, tea, etc.)	12	3
Forestry	5	8
Veterinary science	9	5
Oceanography	1	1
Geography	1	–
Engineering and industrial	5	6
Medicine, health	10	16
Natural history, wildlife	4	6

LANGUAGES OF SCIENCE

An immediate effect of the spread of science has been to multiply the languages of publication. This has been a further extension of the development of the vernaculars for scientific publication. A picture of the situation at the end of the nineteenth century was given by Bolton's survey of the literature of chemistry from 1492 to 1892. He recorded the languages of publication as in table 20.

The "other" languages included a number of European vernaculars, Arabic, and two from the East, Chinese and Japanese. No other non-European language was represented.

Table 20 Languages of Chemistry, 1492–1892

Language	*Books*	*Journals*
German	3,072	195
English	1,732	125
French	1,563	80
Italian	450	29
Latin	403	–
Dutch	277	11
Russian	219	1
Spanish	174	–
Danish	84	6
Others	142	23

More recently, reports of scientific and technical work began to be published in many more languages. However, the number of publications does not correlate well with the numbers of native speakers of the different languages. Table 21 gives the results of two surveys of journal titles: the first, reported by Wood, is an analysis of titles received in 1966 at the British National Lending Library for Science and Technology; the second, reported by Mikhailov, analyzes titles abstracted in 1974 at the Soviet Institute for Scientific Information. The third column is from Crystal, giving the numbers of native speakers of each language as a proportion of the world population (all figures are percentages).

Thus English, only spoken as a native tongue by 8 percent of the world, accounted for between 29 percent and 46 percent of the world's scientific

Table 21 Language Distributions (Percentages)

Language	*Wood*	*Mikhailov*	*Crystal*
English	46	29	8.0
Russian, etc.	14	15	3.5
German	10	13	2.3
French	9	8	1.6
Japanese	4	6	2.8
Spanish	3	2	5.7
Italian	3	3	1.4
Portuguese	2	2	3.1
Polish	2	3	*
Czech, Slovak	2	3	*
Others, including	5	16	71.6
Scandinavian		5	*
Serbo-Croat		3	*
Hungarian		2	*
Romanian		1	*
Others, including		5	
Chinese			23.0
Hindi			4.6
Arabic			3.5
Bengali			3.5
Panjabi			1.6
Javanese			1.5
Bihari			1.5
All others			32.4

*included in "all others" in this column

Sources: Wood (1967), Mikhailov (1984), Crystal (1987)

journals. It should be noted, however, that countries in which English is an official language, where literate people might be expected to know it, account for 32 percent of the world's population. In such countries (e.g., India) scientific publication is often in English. If we take this into account, the predominance of English becomes more understandable.

But this is not the whole story. English, Russian, German and French dominated scientific publication, and many languages contributed far less in relation to the number of their native speakers. The biggest differences were in the lower part of the table: the group of languages from Chinese to Bihari, accounting for 39 percent of the world's population, produced less than 5 percent of the world's science-technology journal titles.

INTERNATIONALISM IN SCIENCE

The many informal links between scientists in different countries, which we have noted earlier, have increasingly been supplemented by formal institutions. International congresses first became a regular feature of the scientific scene in the nineteenth century, though there were a few even earlier: an international gathering of metallurgists met in 1786 at Skleno (Germany), and a congress on metric standards was held in Paris in 1798. The first of the modern international congresses was probably that on statistics, held in Brussels in 1853, organized by Adolphe Quetelet, which met thereafter every two or three years. It was followed by many other series, for example those in table 22.

Table 22 International Congresses, Starting Dates

ophthalmology, 1857
chemistry, 1860
veterinary arts, 1863
botany, 1864
pharmacy, 1865
geography, 1871
meteorology, 1873
geology, 1878
physiology, 1889
mathematics, 1897

A congress on the history of science was first held in Rome in 1903. An Académie Internationale d'Histoire des Sciences was created in 1928 by Aldo Mieli and others, and this has organized international congresses.

An early attempt to integrate scientific activity internationally was the International Association of Academies set up in Paris in 1900. After the First World War an International Research Council (IRC) was instituted (1919). Each country was advised to set up or recognize a central scientific body to represent the country in the council. An International Union was organized in each major field of science to coordinate and develop scientific activities (e.g., for astronomy, geodesy and geophysics, chemistry, radio, pure and applied physics, biology, geography, crystallography). In 1932 the IRC was renamed the International Council of Scientific Unions (ICSU). To deal with problems concerning more than one Union, Joint Commissions were established by ICSU at various times, for example on oceanography, radioactivity and spectroscopy. International data projects were developed, for example, on x-ray crystallography, on the stability of metal complexes, on thermodynamic tables.

In 1950 there was set up the Union of International Engineering Organizations (UATI), which had as its members a number of international associations and commissions—e.g. on hydraulics, structural engineering, irrigation, electric systems, welding, roads, testing of materials—and the World Power Conference, which had first met in 1924. Under the auspices of the United Nations, many international agencies were set up, such as UNESCO, the International Atomic Energy Agency, the World Meteorological Organization, the Food and Agriculture Organization, and the World Health Organization.

Based on 1967 data, Cherry prepared a graph, figure 9, showing how the number of international organizations (not only scientific) had increased during the twentieth century up to that time. Since then, the number has certainly doubled.

The coming together of scientists from various countries into an international project is also not wholly a twentieth-century phenomenon, although it has grown considerably during this period. From 1829 on, under the auspices of the Berlin Academy, a number of German, Italian and Danish astronomers helped to draw up star catalogues. In 1864 an international Stazione Zoologica was set up in Naples by the German naturalist Anton Dohrn to study Mediterranean marine biology. The International Bureau of Weights and Measures was formed in France in 1875. An international network of observatories was created to study variations in the earth's

Figure 9 Growth of International Organizations

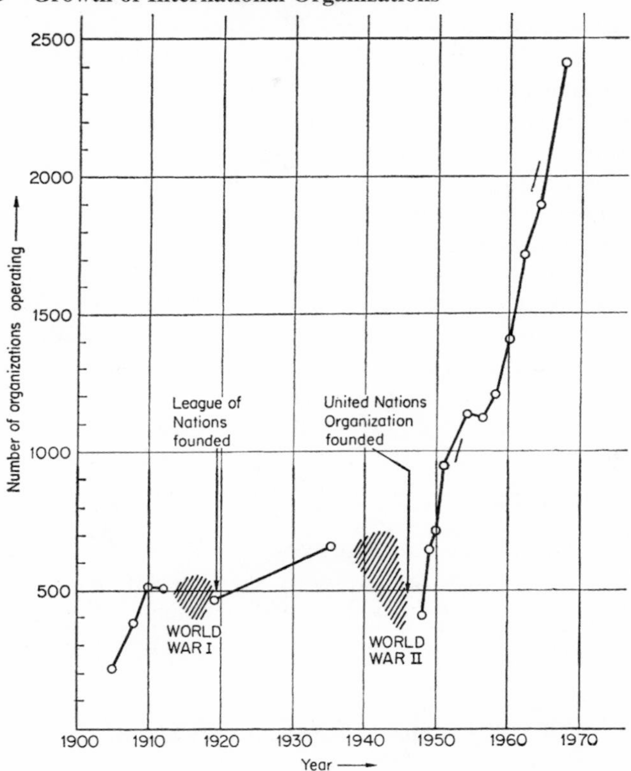

magnetic field. Similar international efforts were made to collect weather data.

"International years," during which cooperative research was undertaken, have been a feature of astronomical and geophysical science—e.g., the International Polar Year (1882–83), the International Geophysical Year (IGY, 1957–58), the International Year of the Quiet Sun (IYQS, 1964 on), the International Space Year (1992). In the IGY, for example, about sixty countries took part, taking observations over eighteen months at hundreds of research stations throughout the world. To cope with the data generated, three World Data Centers were established (in the United States and Soviet Union and one distributed by discipline in Australia, Japan and Europe). The centers continued thereafter to collect other data, for example that resulting from the IYQS. UNESCO initiated a number of international programs, on such themes as Man and Biosphere, Geological Correlation, Hydrology, the Lithosphere, Global Environmental Monitoring.

In Europe, multinational research-and-development centers set up include the European Laboratory for Particle Physics (CERN, 1954), the European Space Agency (ESRO, 1964) and the European Molecular Biology Laboratory (1974).

A considerable number of cooperative research agencies in the field of agriculture were formed in which African countries participated. France took a leading part in this activity, with a center for agronomic research (CIRAD, 1970), and a series of Instituts de Recherches, set up from 1936 on, covering topics such as cocoa and coffee, cotton, vegetable oils, rubber and fruits. There were created cooperative locust-research organizations in Ethiopia (1962) and Zambia (1970). Initiatives related to the so-called "green revolution" in agriculture should also be noted: in the 1940s there was established in Mexico the Centro International de Majoramento de Maize y Trigo, and in 1960 the International Rice Research Institute in the Philippines.

All these international and cooperative activities were in themselves, of course, modes of communication among scientists, but they also generated scientific records that had themselves to be communicated—congress proceedings, recommendations, regulations, research reports, and a vast mass of data.

Parallel to the growth of internationalism in science there was a corresponding development in efforts to coordinate scientific and technical information. In 1895, two Belgian lawyers, Paul Otlet and Henri la Fontaine, founded in Brussels an Institut Internationale de Bibliographie, with the aim of creating a world clearinghouse for bibliographical information. Among their products were the first edition of the Universal Decimal Classification (1905) and a voluminous *Traité de documentation* (1934). A world congress on universal documentation was held in Paris in 1937. The Royal Society of London in 1948 held a scientific-information conference, with attendance from the British Commonwealth and from the United States, that ranged widely over the problems of scientific communication, and this was followed up ten years later by an international conference on scientific information in Washington, D.C., sponsored by the U.S. Academy of Sciences. In the 1960s, UNESCO and ICSU set up the project UNISIST, to explore the feasibility of a world science-information system, and this reported in 1971. The UNISIST program laid emphasis on the promotion of (a) tools to aid the interconnection of information systems, such as standards; (b) assistance in the strengthening of information provision in developing countries; and (c) research into the problems of communication in science.

DEVELOPMENTS IN SCIENTIFIC PUBLICATION

The amount of scientific and technical literature that will have been published during the twentieth century is likely to amount to fifty times that produced during the whole preceding period up to 1900. Any individual publications mentioned below will therefore be only a tiny sampling of what has been produced. Scientific books and journals continued to proliferate relentlessly. Many scientific and technical periodicals were still produced by societies, but other kinds of publisher became important: academic institutions, research institutes, government agencies, industrial manufacturers and, above all, commercial publishers.

Complaints by scientists about delays in the publication of papers, by whatever journals were currently available, have been perennial. News journals carrying brief announcements of recently completed research tried to offset the delays—e.g., *Nature* (1870), *Science* (1883), *Naturwissenschaften* (1913). By mid-twentieth century, journals of general coverage such as these were deemed inadequate to cope with the problem of early announcement, and a new breed of "letters journals" began to appear in the 1960s, for example in electronics, chemistry, organic chemistry, chemical physics, physics. Informal reporting of research efforts, particularly in newly emerging specialties, more recently took the form of newsletters, often produced in near-print. Some examples were *Drosophila Information Service*, *Superconductivity News*, *Arthropod-Borne Virus Information Exchange*, and *Newsletter on Analysis of Astronomic Spectra*.

Specialization of journals continued to increase. As a small example of this, appendix C presents a chronological survey of the appearance of journals related to physical chemistry. At first this subject was part of general chemistry, as designated in the appendix, starting with *Journal fur praktische Chemie* in 1828 and spreading out over many countries. Then in 1887 the first explicitly named journal in physical chemistry appeared—*Zeitschrift fur physikalische Chemie*—to be followed by others. In 1902 various subdivisions of physical chemistry began to give rise to their own journals. General-chemistry journals continued to be founded in ever more diverse countries—the latest listed being the *Chinese Journal of Chemistry* in 1983—as well as many journals of physical chemistry. Journals covering subdivisions of physical chemistry included those on electrochemistry, colloids, catalysis, spectroscopy, reaction kinetics, and quantum chemistry; few of these were published in countries outside Europe and the United States.

During the century there were started a number of major abstracting and indexing journals with wide coverage, such as those in table 23.

Many new, specialized abstracts journals also appeared—for example, within chemistry there were produced abstracts journals on catalysis, adhesives, macromolecules and polymers, chromatography and other subjects. From the journal *Chemical Abstracts* there have been produced since 1976 a large series of *CA Selects*, each covering a specific topic from the parent journal. Some abstracts were produced that were restricted in coverage to publications from a particular country or region, e.g., *Indian Science Abstracts*, and *Bibliografia Brasiliera.*

The twentieth century has seen the development—or major expansion—of forms of scientific publication that were not so prominent earlier. We will briefly review a number of these: conference proceedings, reviews of progress, patents, standards, trade literature, data compendia, technical reports, translations, dictionaries, and guides to the literature.

International congresses, discussed earlier, are not the only kinds of gathering that have given rise to published proceedings. There has been a

Table 23 Abstracting and Indexing Journals

Start date	*Title*
1907	*Chemical Abstracts*
1913	*Industrial Arts Index*
1916	*Agricultural Index*
1926	*British Chemical Abstracts*
1926	*Biological Abstracts*
1929	*Geological Abstracts*
1938	*Dissertation Abstracts*
1940	*Mathematical Reviews*
1940	*Bulletin analytique, CNRS*
1942	*Bibliography of Agriculture*
1946	*Excerpta medica*
1947	*Nuclear Science Abstracts*
1953	*Referativnyi zhurnal*
1959	*Geoscience Abstracts*
1961	*Science Citation Index*
1963	*International Aerospace Abstracts*

plethora of conferences, conventions, symposia, seminars, workshops, round tables, clinics, institutes, colloquia, study groups, summer schools, training sessions and just plain meetings in every field of science and technology—to say nothing of technical committees and scientific commissions. Attendance at a meeting might be restricted, e.g., to committee members, or to members of a university department, or to a specially invited group of scientists; or it might be open to anyone who paid the inevitable fee. The scope of attendees might be local or regional, national or international. Each kind of meeting could generate documentation, ranging from handouts circulated at a training session to printed multivolume proceedings, or even audiotape or videotape recordings.

The value to the participant of any kind of meeting was not only its formal content, including the scientific and technical information presented, and the general discussion provoked. It offered also a chance of informal contact with colleagues or newly met professionals. Despite all the documentary (and later electronic) means of communication that existed, scientists found that this face-to-face informal contact was both stimulating and necessary. It is no wonder that the British Library, which made a special effort to acquire conference proceedings worldwide, should in 1990 have been recording eighteen thousand items per year—and this did not take into account the tens or perhaps hundreds of thousands of meetings each year that did not result in publicly available documents.

Reviews of progress were certainly not a new phenomenon—in an earlier section a graphic description was given of Berzelius in the 1820s compiling *Jahresberichte*. But there was a considerable expansion during the twentieth century of annual review series, and in many surveys asking scientists what literature services they would like, a call for more reviews was heard. The demand is indeed inexhaustible—there is never an adequate review available of the latest "hot topic." Attempts to meet the need were often made at scientific meetings—for example, in 1968 Garvey estimated that at a sample of U.S. national science and engineering meetings, from one quarter to one half of the program was devoted to state-of-the-art and review presentations.

Patents can even less be described as a new development, and their expansion during the nineteenth century has already been discussed. By 1990, about three-quarters of a million patents were being issued annually throughout the world, and the total volume of patents exceeded 27 million. A World Intellectual Property Organization was set up by the United Nations in 1967, and it drew up in 1970 a Patent Cooperation Treaty that was signed by many countries: the filing of a patent application in one country could then automatically lead to the application being filed in other designated countries. A European Patent Convention was established in 1973, permitting an application in one European country to lead to the granting of a Europe-wide patent.

Standard specifications of products and processes have become of great importance to industry: to simplify the growing variety of products and procedures, to communicate information, to improve economy of production, to set safety standards, and to eliminate trade barriers between countries. The first national standards body, known as the Engineering Standards Association, was set up in Britain in 1901, becoming the British Standards Institution in 1929. The Deutsches Institut für Normung was formed in 1917, and the French Association Française de Normalisation in 1926. In the United States a number of standardizing organizations were set up, such as the American Society for Testing Materials (1902), and the American Standards Association (1918, now the American National Standards Institute). An International Electrotechnical Commission was formed in 1906, and this collaborated with the International Organization for Standardization (ISO) set up in 1947. A Joint European Standards Institution was created in 1985. There were many other international and national organizations that issued standards, acceptable in varying degree to different industrial groups.

Trade literature produced by industrial firms has also grown in importance. Trade catalogues appeared long before the twentieth century—for example, a guide to American trade catalogues 1744–1900 was published in 1960, and there are library collections going back to the 1850s. House journals published by leading manufacturers such as IBM, Marconi, and ICI often contained valuable scientific and technical material.

Compendia comprehensively covering a field of knowledge are a narrowing and deepening of the encyclopedias that we have noted in the eighteenth century. Their production began towards the end of the nineteenth century—particularly in the form of German *Handbücher*, systematic accounts of what was known, with extensive references to the literature, for example, those in table 24.

Many similarly comprehensive treatises, manuals and handbooks have since been published. Some examples are shown in table 25.

Alphabetically arranged specialist encyclopedias also proliferated during the century, some major examples being shown in table 26.

Data compendia, concentrating particularly though not exclusively on quantitative information in a subject field, also arose in the nineteenth century. The German chemist Leopold Gmelin first published a *Lehrbuch*

Table 24 Some Early *Handbücher*

Handbuch der Ingenieurwissenschaften, 1879 on
Durm and Ende, *Handbuch der Architektur*, 1892–1907
Wolf, *Handbuch der Astronomie*, 1890–93
Valentiner, *Handworterbuch der Astronomie*, 1897–1902
Kayser, *Handbuch der Spektroscopie*, 1901–12
Korscheldt, *Handworterbuch der Naturwissenschaften*, 1912
Abderhalden, *Handbuch der biologischen Methoden*, 1920–39
Oppenheimer, *Handbuch der Biochimie*, 1924–7
von Bethe, *Handbuch der Physiologie*, 1925–32
Wien and Harms, *Handbuch der experimentelle Physik*, 1926
Flugge, *Handbuch der Physik*, 1926–9

Table 25 Some Comprehensive Treatises

Glazebrook, *Dictionary of Applied Physics*, 1922–23
Mellor, *Comprehensive Treatise on Inorganic and Theoretical Chemistry*, 1922–37
Hyman, *The Invertebrates*, 1940–59
Partington, *Treatise on Physical Chemistry*, 1949–54
Kolthoff, *Treatise on Analytical Chemistry*, 1958–76
Florkin and Stotz, *Comprehensive Biochemistry*, 1962 on
Ainsworth and Sussman, *The Fungi*, 1965–73

Table 26 Some Alphabetical Encyclopedias

Kirk-Othmer, *Encyclopedia of Chemical Technology*, 1947–60
Thewlis, *Encyclopedic Dictionary of Physics*, 1961–65
Mark, *Encyclopedia of Polymer Science/Technology*, 1964–72
Encyclopedia of Industrial Chemical Analysis, 1966–74

der Chemie in 1817, and this later grew into the multivolume *Handbuch der anorganische Chemie* produced by the Gmelin Institut. Friedrich Beilstein began to publish his *Handbuch der organischen Chemie* in 1918. The *Zahlenwerte und Funktionen aus Physik, Chemie, Astronomie, Geophysik, Technik* were first produced by Landolt and Börnstein in 1883; a new series began in 1961. The *Tables annuelles de constantes et données numériques* were published in France from 1912 on for a number of years. After studying these, in 1919 the International Union of Pure and Applied Chemistry assigned to the United States the responsibility for producing a data compilation in physical science; the International Research Council (now ICSU) sponsored the work, and as a result the *International Critical Tables of Numerical Data in Physics, Chemistry and Technology,* edited by Washburn, began to appear in 1926. Many specialized data compendia have since been produced, such as those in table 27.

Technical reports, most often arising from projects funded by government agencies, began to contribute importantly to scientific communication after the Second World War, although earlier instances occurred. For example, geological surveys, even early in the nineteenth century, were issued in report form. The report series of the U.K. Aeronautical Research Council started in 1909; that of the U.S. National Advisory Committee on Aeronautics in 1915.

The ending of the 1939–45 war itself inaugurated a big report problem: the Allies sent into Germany, in the wake of the armies, teams of investigators charged with the duty of seizing from German industry as much technical documentation as could be found. Both in the United Kingdom and the United States, agencies were established to review, reproduce and distribute this material for the benefit of their own nations. About three thousand summary reports in English were prepared in the United Kingdom and offered for sale. A vast mass of original German documents remained. Summaries of these were made (by 1948, seventy-six thousand summaries had been produced in the United Kingdom) and offered free to industry, who could then inspect the original document or purchase a photocopy.

Since then, a great generator of technical reports has been the United States, with material emanating from the Atomic Energy Commission, the National Aeronautics and Space Administration, and many other government sources. Although the existence of such "gray" literature was announced in various bibliographical serials, and depository libraries for it

Table 27 Some Data Compendia

Standard Handbook for Electrical Engineers, 1907
Handbook of Chemistry and Physics, 1913
Machinery's Handbook, 1913
Standard Handbook for Mechanical Engineers, 1916
Chemical Engineer's Handbook, 1934
Handbook of Biological Data, 1956
Handbook of Toxicology, 1956–9
Nuclear Handbook, 1958
Biochemist's Handbook, 1961
Computer Handbook, 1962

were designated, access was never easy. Various U.S. agencies were at different times responsible for providing announcements and giving access to technical reports, such as the National Technical Information Service (NTIS), which by 1990 covered about 1.5 million items. In 1978 a European Association for Grey Literature Exploitation was set up, in which the national information and document supply agencies of a number of European Community countries participated, to improve access to reports and other informally published documents (project SIGLE). By 1990, data on 180,000 European documents had been collected.

Another kind of "grey" document whose production has expanded during the last half century is the translated scientific or technical paper. It was estimated in 1990 that material of this kind totaling 4 billion words was translated annually in the United States alone. Attempts to identify, collect, and make available such translated papers started in the early 1950s—at ASLIB in London and the Special Libraries Association in the United States. In 1961 a European (later International) Translations Center was set up at the technological university of Delft, and the British Library began to make a comprehensive Anglo-American collection. In the United States, a National Translations Center was set up in the John Crerar library, Chicago, and was later transferred to the Library of Congress in Washington.

Soviet success in space activities led in the 1950s to a flurry of interest in other countries about the level of Soviet scientific and technical achievement. U.S. and U.K. government representatives decided that cover-to-cover translations into English of key Soviet journals should be sponsored. Many hundreds of such journals were being translated by the 1970s, although the number decreased somewhat in later years.

Translation requires the availability of multilingual dictionaries, many of which have been produced. But apart from this, the size and complexity of scientific vocabulary has given rise to considerable activity in the production of monolingual terminologies of various kinds. Dictionaries, both general and special, are of long standing. A work such as the *McGraw-Hill Dictionary of Scientific and Technical Terms* might contain over a hundred thousand entries, but this covered only a fraction of the hundreds of thousands of known plants and animals, and the millions of chemical substances. Many specialist glossaries have been produced in attempts to standardize the terminology of particular subject fields, for example the *Glossary of Geology* of the American Geological Institute, 1972. A further form of terminology developed since the 1960s has been the thesaurus, a list of standard terms for use in indexing. In 1971 there was established in Vienna under UNESCO auspices an International Information Center for

Terminology (INFOTERM), to coordinate terminological activity throughout the world.

The range of different publications becoming available also led to the appearance of specialized "guides to the literature" of particular subjects. Early examples were the books by Felix Müller, *Führer durch die mathematische Literatur* (1909); by A.B. Eason, *Where to Seek for Scientific Facts* (1924); and by E.J. Crane and A.M. Patterson, *Guide to the Literature of Chemistry* (1927). Many others followed.

LIBRARIES AND INFORMATION CENTERS

Throughout the twentieth century libraries serving science and technology continued to be formed—organized by societies, academic institutions, research institutes, government departments and agencies, industrial firms and, of course, by the public library system that had been solidly established during the previous century. Throughout the twentieth century, also, the concept and practice of library service has altered. The nineteenth-century librarian was often a scholar (or would-be scholar), interested in contributing, even if only in a modest way, to the development or history of the subject covered by the library. Increasingly, there evolved the concept of an "information" scientist—that is, a scientist whose specialty was that of locating, collecting, organizing, disseminating and processing information. The services provided by such an information scientist have grown steadily more dynamic, more closely tuned to the needs of the particular group served, more innovative in finding ways to meet those needs.

Undoubtedly it was the formation of industrial libraries in the late nineteenth and early twentieth centuries—noted in an earlier section—that triggered this development. As well as performing the normal functions of document collection and loan, industrial librarians began to disseminate information by means of circulating issues of periodicals, issuing bulletins giving titles or even abstracts of newly received papers, establishing detailed indexes to locally held material, undertaking literature searches and translations, actively cooperating with other libraries in document provision. Service to industry even spread to the more active public libraries in industrial towns. Let us take the United Kingdom as an example: the Glasgow public library established a separate commercial section in 1916; in 1934, Sheffield formed a separate technical library, and followed this up in 1935 by setting up a cooperative scheme for the exchange of publications between the industrial, university and public libraries of the city. Even more significantly for the future, in 1926 a major government library, the

Science Museum Library in London, then under the direction of S.C. Bradford, began to lend books and journals to other libraries, a facility of which industrial libraries made good use. To aid interlibrary cooperation, a *World List of Scientific Periodicals* was produced in 1925–27, giving their locations in British libraries, and ASLIB (see below) in 1928 published the first guide to the specialized information resources of Great Britain. In due course (1962) a National Lending Library for Science and Technology was set up at Boston Spa in Yorkshire.

In Britain, a contributing factor to the development of new ideas and practices in information provision was the governmental Department of Scientific and Industrial Research (DSIR). This was created in 1916, and became responsible for a number of research laboratories. More significantly, the department helped to fund and guide an increasing number of industrial research associations (ten by 1920, forty by 1950), part of whose remit was to act as information centers for the industries served.

The new trends led to the formation in the United States of a Special Libraries Association (1909), and in the United Kingdom the Association for Information Management (ASLIB) was created (1924). The continuing evolution of the "information" concept led in due course to two further organizations, an American Documentation Institute in 1937 (later to become the American Society for Information Science) and the U.K. Institute of Information Scientists. An International Federation for Documentation (FID) was formed in 1931, a direct descendant of Institut Internationale de Bibliographie, founded in Brussels by Paul Otlet and Henri la Fontaine in 1895.

By 1950, World War II had come and gone, making great calls on science and technology, and demonstrating the need for information to back up scientific and technical development. A leading American science administrator, Vannevar Bush, in 1945 stated that "scientific publication has been extended far beyond our present ability to make real use of the record. . . . The means we use for threading through the consequent maze to the momentarily important item is the same as was used in the days of square-rigged ships." In the following year, the first electronic digital computer was constructed. In the same year, UNESCO was founded, and three years later held an international conference on science abstracting. In 1948, the Royal Society of London organized a Scientific Information Conference that had a great impact. In 1949, Claude Shannon and Warren Weaver published their influential *Mathematical Theory of Communication*. The stage was set for decisive new advances in scientific communication, and libraries and information services began to respond to the new situation.

Federal government agencies in the United States (such as the Library of Congress, the National Library of Medicine, the National Agricultural

Library, the Department of Commerce, the Atomic Energy Commission, the Patent Office and others) began to exert an important influence on the development of scientific-communication processes. Their activities included the production of abstracting or indexing serials, the establishment of regional information centers, and experimentation with new information technologies.

A further development, particularly in the United States, was to advance from the concept of "information center" to that of "information analysis center"—an organization that not only collected, stored, organized and disseminated information, but also critically evaluated and collated it, disseminating the results of its analytical work. Some examples of this existed even in the nineteenth century, in the form of the major environmental-data programs (geology, geodesy, weather, etc.). During the 1950s and 1960s, a hundred or more such centers were formed, usually with U.S. government funding, in such diverse subject areas as radiation effects, chemical propellants, shock and vibration, oceanographic data, air pollution, nondestructive testing, reliability analysis, cancer chemotherapy, Parkinson's disease, and ceramics.

Information analysis centers of this kind played an important part in the compilation of numerical data, such as that provided in the data compendia described earlier. A National Standard Reference Data Center was set up in 1963 by the U.S. National Bureau of Standards to collect numerical data of scientific interest, critically evaluate it, and compile it in orderly format.

At the other end of the scale from these very specific information activities, during the 1950s and 1960s a number of national scientific and technical documentation or information centers were set up, sometimes under UNESCO auspices, e.g., in India (1952), Czechoslovakia (1952), Poland (1953), Uruguay (1953), Brazil (1954), Sri Lanka (1955), China (1956), Pakistan (1956), Bulgaria (1959), Thailand (1961), Cuba (1963), Belgium (1964), Indonesia (1965), Morocco (1966), Bolivia (1967), Iran (1968).

Increasingly, the further evolution of libraries, information centers, information analysis centers, and documentation centers became dependent on the introduction of new technology, and it is to this topic that we must now turn.

COMPUTERS FOR PROCESSING INFORMATION

Although computers and their application to information handling are such a recent development, it is of interest to look back a little at the growth of the idea of machine manipulation of concepts. The idea of a "logic machine" is far from new. Probably the first advocate of a mechanism to aid the logical combination of ideas was the thirteenth-century Ramon Lull of Majorca, in his *Ars magna*. In essence his method was to arrange two sets of concepts around the edges of two concentric circles, then to rotate one circle against the other and consider the numerous combinations of "outer" and "inner" concepts that this rotation created. Jonathan Swift may have had Lull's art in mind when he described the academy of Laputa in *Gulliver's Travels* (1726). There Gulliver was shown a frame containing hundreds of wooden cubes linked by wires. On each side of each cube was a word. Handles could rotate the cubes to produce random combinations of words. By this means, the professor of the academy averred, "one may write books in philosophy, poetry, politics, law, mathematics and theology, without the least assistance from genius or study." In this same eighteenth century, Charles Stanhope built a "demonstrator" that could be used to perform logical operations. More effective than this was William Jevons' logic machine of 1869—the first machine powerful enough to solve a complicated set of syllogisms faster than an unaided human.

Much superior in design—but aimed at numerical rather than logical computation—was the analytic engine on which Charles Babbage worked from 1833 to 1871. It had two features that were to be fundamental for future computers. First, it was a general machine, not confined to a specific task. By altering its data and instructions, it could be set to perform different tasks. Second, the data and instructions were fed to the engine in "machine-readable" form—in fact, in the form of cards in which holes were punched. Cards of this kind had been used in 1805 by Joseph-Marie Jacquard to represent patterns to be created by a weaving loom (they are still in use in China to create patterned silk fabrics). In 1890 Herman Hollerith used such cards in a tabulating machine to analyze U.S. census figures, later forming a company that has since become IBM.

Meanwhile, the use of symbols in logic had been developed by Augustus de Morgan (1838) and George Boole (*Laws of Thought*, 1854). Claude Shannon in 1937 showed how electrical networks could represent logical operations, and in the same year, Alan Turing published a seminal paper on the nature of computing operations. Unknown outside Germany, at about the same time Konrad Zuse developed a mechanical calculator using Babbage's ideas, later employing electromagnetic and even electronic components. In the United States from 1937 on Howard Aiken worked on an electromechanical calculator that was later built by IBM for use by the U.S. Navy. In 1939, John Mauchly and J. Presper Eckert led a team that developed the first electronic digital computer, the ENIAC, completed in 1946.

Maurice Wilkes built on their work to complete the EDSAC computer three years later. In 1951, the Remington Rand firm produced the first commercially available computer, UNIVAC-1, and IBM soon began to market machines. In England, the first commercial machine was built by Ferranti.

It soon became apparent that the digital computer was essentially a device to manipulate symbols (mathematical computation and logical operations were only two examples of this). If we knew how to represent physical or mental elements and processes symbolically, and could write a set of instructions on how the symbolic data was to be manipulated, then the operations could be performed by computer. As Ada Lovelace (mathematician and colleague of Babbage) said in 1842, "the analytical engine . . . can do whatever we know how to order it to perform." So in due course methods were devised to enable the computer to handle not only numbers and logical propositions but also word indexes to be searched, text to be scanned, language to be processed, images to be manipulated, and much more.

Early computers were huge machines—the ENIAC covered fifteen hundred square feet of floor space, and weighed thirty tons. This was mainly because the basic network components—vacuum tube valves, first invented by Lee de Forest in 1907—were each the size of a light bulb, interconnected by a tangle of wires (ENIAC used nineteen hundred such valves). In 1948, in place of the valve the much smaller transistor was developed by John Bardeen, William Shockley and Walter Brattain at Bell Laboratories, and more compact, more reliable and faster computers could be built using this device. In the 1960s, the invention of integrated circuits greatly enhanced the process of miniaturization—hundreds of thousands of transistor elements (and their interconnections) could be deposited photochemically on a silicon chip the size of a small postage stamp.

Meanwhile, programming (the method of writing instructions for the computer) progressed from laboriously spelling out a series of elementary computer operations, to the use of sophisticated and flexible "high level" programming languages—first FORTRAN in 1956, followed by COBOL in 1959, BASIC in 1965, and a host of others.

The first microprocessor—from which a computer system small enough and cheap enough to be used as a personal tool could be constructed—was developed by M.E. Hoff of Intel in 1971. It consisted of about two thousand transistors on one chip, and was roughly as powerful as the ENIAC. Its first competitor, from Motorola, arrived in 1973. The Altair microcomputer of 1975 was the first commercial machine to use the Intel chip. It included the BASIC programming language, in a version developed by

Bill Gates, later founder of Microsoft. Since then, microcomputers have become ubiquitous, as, indeed, have computers of every variety. The spread of use was facilitated by continuing simplification of ways for humans to interact with them. Punched-card input gave way to keyboards, later supplemented by devices such as the mouse, the graphics tablet, and optical character recognition. From the cathode ray tube—a nineteenth-century invention subsequently used in television—the video display (VDU) screen was developed for computer output. Screen display has become ever more flexible and sophisticated. For hard-copy output, printers too have continuously evolved.

The myriad contributions of the computer to science, medicine and technology are not the theme of this book, but particular mention should be made of its impact on printing and publishing. The use of computers in scientific communication was made possible by several other developments. First, the computer capacity to store data was enormously extended. Vast quantities of information could be put into "machine-readable" form on magnetic media, ready for access by a retrieval program. Second, from the 1960s specialized software programs were developed that could efficiently perform retrieval from these huge files, and permit simultaneous multiple access to them ("time-sharing"). Third, facilities for linking computers together were also greatly enhanced. Local networks of computers could be constructed, using appropriate wiring. Remote computers could be accessed via the lines and radio links already available in the telephone system. Fourth, the telecommunications system itself became more powerful and flexible, using new types of lines (e.g., optical fibers, first developed in 1955) and new ways of transmitting electronic messages (e.g., package switching). A new physical path for such messages became available with the launching of communications satellites (e.g., Telstar, 1962; the INTELSAT system, 1968 on). Direct overseas dialing began to be introduced in 1971. The development of messaging systems opened the way to electronic mail from one person to another, or to all persons on a mailing list; this in turn made possible the distribution of bulletins, newsletters or journals electronically, and conferencing via computer links. Control of all these communication facilities was itself made possible by the use of computers.

DATABASES AS SOURCES OF INFORMATION

Since 1950, computer-database manipulation evolved along two lines that for the most part had little contact, although more recently there has been a rapprochement. The two types of database handled were, on the one hand, structured databases, particularly though not exclusively used for data associated with financial and industrial management, and on the other hand, databases of relatively unstructured text, used for scholarly and technical information of all kinds. In both cases, software to search and manipulate the databases was built both by individual user institutions and by commercial suppliers. The emphasis in this book will be on text databases, but some account will first be given of structured databases.

In the 1950s, computer manufacturers were selling only hardware. The number of operational computers was low—some hundreds, located in major firms and government installations, and a few universities. Software had to be developed by the computer users, and as computers spread further into industry and commerce, companies built files relating to their activities and wrote programs to manipulate them. It was not till the sixties that hardware manufacturers began to develop and offer database management software that could be used for such files. Often a vendor would create a system under contract to a large user. Thus IBM built a program package, DL/1, for North American Aviation as part of the NASA Apollo program. In the early seventies, Honeywell built MDQS as part of the U.S. Worldwide Military Command system. Software suppliers were formed, such as Informatics, who built a package, GIRLS, for the U.S. Air Force, and System Development Corporation created TDMS for the U.S. Defense Department. Some users developed systems in-house, initially for their own use, thus Goodrich Chemical Company built IDMS, and later marketed it. Other software houses developed programs designed from the start to be offered commercially, such as the TOTAL system of Cincom Systems and RAMIS built by Mathematica.

Over the years, many such programs for handling structured data—now known generally as database management systems (DBMS)—were created, some for large ("mainframe") computers, some for medium size ("mini") computers, some for microcomputers, others with versions usable on several sizes of machine. Such software was helpfully used for information that consisted of numerical fields (e.g., values of a variety of physical properties of chemical substances), or of fields with short, unambiguous descriptive terms, taken from a relatively brief predefined set (e.g., technical terms in taxonomy). But there was a great deal of scientific and technical information that did not conform to this pattern. It was primarily in textual form, there was no restriction on the kinds and number of words used, there was much potential ambiguity of vocabulary. For such information, text management software (TMS) was required, and it is on this that we will now concentrate. During the late 1980s, a number of commercial vendors began to develop packages that combined the facilities of data management and text management, for example, ORACLE and BASIS-PLUS (created by Information Dimensions).

The same institutions of the 1950s that were considered above also typically had documentation departments. The idea of making use of the newly available computers for document processing occurred to some users. Most concentrated on records containing some form of indexed bibliographic reference. Probably the earliest experiment in computerized reference retrieval was that of P.R. Bagley, a student at the Massachusetts Institute of Technology, who in 1951 concluded, however, that a general-purpose computer was not an effective tool for this purpose.

Disappointingly slow search was also reported in 1957 by R.H. Bracken and H.E. Tillitt at the U.S. Naval Ordnance Test Station. In both cases, bibliographic data was stored on magnetic tape, and the whole tape had to be scanned serially to locate entries that matched the search criteria. There was another problem. At that time, computers were being used mainly for heavy numerical computations. The workload was such that wherever possible short jobs were batched: as a consequence, even if an experimental retrieval program was up on the computer, a search query would have to wait until a sufficient number of other queries could be batched together to make up an economical run. All in all, a computer installation could not at that time provide quick answers to queries.

An inventive IBM engineer, Hans Peter Luhn, saw that information on tape could be used in other ways. In 1958, he put forward two proposals. The first was to record references on tape and use the computer to generate an index that could be printed out, replicated and distributed. The particular form chosen was the so-called KWIC index, in which each significant word in a document title was used as an index entry (showing it in its context in the whole title), and these entries were alphabetically arranged. The first regular publication using this technique was *Chemical Titles* from the Chemical Abstracts organization.

The second idea put forward by Luhn in 1958 was "selective dissemination of information" (SDI). Each month (or other convenient interval), indexed bibliographic references would be put onto tape; each of a group of (perhaps hundreds of) users would supply a set of search criteria; the combined sets would be batched and run against the monthly tape; items selected for each query set would be printed out and dispatched to the users. The system was implemented by IBM in 1959. In 1962 the producers of *Chemical Abstracts* decided to offer monthly tapes containing their indexed abstracts for sale to institutions that could use an in-house SDI program to provide a service to their staff. By 1963, the Ames laboratory of the U.S. Atomic Energy Commission had designed an SDI system that could cope with a variety of tapes from different producers. In 1965, the U.S. Institute of Scientific Information (ISI) began to offer a commercial service, providing subscribers with regular selected printouts from their *Science Citation Index* tapes.

Both of Luhn's ideas became very popular. Many information centers produced KWIC (some called them "KWIC and dirty") indexes to specialized bibliographies. Many industrial firms, government agencies and academic institutions developed or purchased SDI software, bought tapes from producers, and offered SDI services. In a 1969 survey of thirty-two U.S. industrial and governmental in-house SDI services, the number of users per service ranged from fifty to over two thousand, the average being nearer the lower value. By 1971, a survey by ASLIB identified about ninety agencies that were regularly producing tapes containing scientific information, mostly in the form of bibliographic references, but also including some data. Many of the agencies were selling or leasing the tapes to customers. Some of the agencies followed ISI in also offering a commercial SDI service.

The agencies in question were mainly institutions producing serial bibliographies—and, in fact, the tapes were initially used in the process of producing and printing those bibliographies. Computer-aided production of abstracts journals had begun in the early 1960s, with *Scientific and Technical Aerospace Reports* from NASA, *Index medicus* from the U.S. National Library of Medicine (NLM), and *Chemical-Biological Activities* from Chemical Abstracts.

Parallel to this, two other lines of development were proceeding. In 1958, IBM introduced the magnetic disk, a form of storage in which it was possible to move readily from any one point to another, and this opened up the possibility of creating a "random access" file structure (comparable to the index and numbered pages of a printed book) that could be searched much more speedily than by serially scanning magnetic tapes. In 1963, an operational computer time-sharing system was implemented at the Massachusetts Institute of Technology (project MAC). With time-sharing, a number of people could use the computer simultaneously, their jobs on the system being interwoven, and batching was no longer necessary. In 1964, Martin Kessler used project MAC to develop an online system to retrieve references from a bibliographic database. By 1965, thirty time-sharing systems were operational in the United States. With these developments, the way was open for fast-response online searching of computer databases.

At about this time, the U.S. aircraft manufacturer Lockheed began developing online retrieval facilities that formed the basis of both RECON, a search system used by NASA, and of DIALOG, which eventually became the major commercial online service. A little later, System Development Corporation (SDC) produced the ORBIT system, a variant of which (ELHILL) was adopted by NLM. The early 1970s saw the development of the commercial online software systems BASIS (Battelle Memorial Institute), STAIRS (IBM), and STATUS (U.K. Atomic Energy Authority). Since then, text-retrieval software has proliferated, for mainframes, minis and microcomputers.

Commercial online services began to be offered from the mid-sixties. Lockheed went public with the DIALOG service, managed by Roger Summit, and SDC with ORBIT, managed by Carlos Cuadra. The NLM used the records created for *Index medicus* to provide the MEDLINE service. At about the same time, the European Space Research Organization organized its online Space Documentation Service, later to become the European Space Agency Information Retrieval Service (ESA/IRS), using RECON software, and headed by N. Isotta. Over the next ten years these services began to mount databases derived from tapes supplied by the major abstracting services, as in table 28.

Online services ("hosts," as they came to be called) were started in countries other than the United States. By 1990, Cuadra was listing in his directory over five hundred hosts, of varying size and importance. A few leading examples in science/technology, apart from those already mentioned, are shown in table 29.

Outside North America and Europe, online services were set up in Japan, Australia, New Zealand, and some Latin American countries. At first,

Table 28 Early Online Databases

Start dates	*Titles*
1966–68	Scientific and Technical Reports (NASA)
1969–70	Nuclear Science Abstracts
	U.S. Government Reports Abstracts
1971–72	CAIN (U.S. National Agricultural Library)
	Chemical Abstracts
1973–74	Compendex (Engineering Index)
	American Petroleum Institute Abstracts
	Georef
	INSPEC (U.K. Electrical Engineers)
	BIOSIS

Table 29 Some Early Online Hosts

BLAISE (British Library)
CAN/OLE (National Research Council of Canada)
DIMDI (Deutsches Inst. Medizinische Dokumentation)
ENDS (European Nuclear Documentation System)
INSERM (Institut National de la Sante, Paris)
Royal Institute of Technology, Stockholm
TELESYSTEMES (PTT, France)
INFOLINE (Pergamon Press)
STN (U.S./German/Japanese partnership)

services such as these had to be accessed directly through terminals connected to the dial-up telephone network. Telecommunication could be expensive, slow and error-prone. Much more effective network services using packet switching techniques to manage connections appeared in the 1970s, such as the U.S. government–sponsored ARPANET, followed by the commercial TYMNET and TELENET. These three originated as U.S. networks, but were later extended to Europe and many other areas; they were used for all remote computer connections, not only for online retrieval.

In Europe itself, the European Commission developed the network EURONET specifically to access retrieval services. In later years, many general-purpose national and regional networks were constructed, such as DATAPAK (covering a number of European countries), PSS (United Kingdom) and the international IPSS, and EURONET was no longer needed. Plans were being made to replace the current type of telephone system with a digital network, ISDN.

In the 1980s, specialized telecommunications networks were set up in many countries to connect university and other academic and research sites, for example, JANET in the United Kingdom, SURFNET in the Netherlands, EARN (the European Academic and Research Network), NSFNET in the United States. International research networks were formed, e.g., HEPNET (high-energy physics research) and the more general system BITNET, which by 1990 was linking thirteen thousand computer centers in thirty-seven countries. Internet, an overarching international network, accessible to all types of users (not just research), was developed: by 1993 it was linking together thousands of wide-area networks that between them interconnected over a million computers worldwide. These networks were used for three main purposes: communication (for example, by electronic mail), distributed computer processing, and access to remote information sources. The developing National Research and Education Network (NREN) in the United States in 1989 defined its long-term goals as:

- to connect every scholar in the world to every other scholar
- to connect to the network all important information resources, computing resources and specialized instruments
- to build databases that are collaboratively and dynamically maintained and contain all that is known on a particular subject
- to create a knowledge management system on the network that will enable scholars to navigate through these resources.

As the use of microcomputers expanded, they began to replace "dumb" terminals as user tools to access hosts. This created the possibility of providing microcomputer software to aid online access—e.g., communications packages that would manage the procedures for establishing a telecommunications link with the host and for downloading search output into the microcomputer; packages for converting downloaded records into other formats; word processors for creating texts into which downloaded data could be inserted; "intelligent" programs that could aid the search process. Online search potentially became more integrated into the other microcomputer-based activities of the scientist and technologist.

The number of available databases grew rapidly. In 1977 Hall described one hundred, and in 1980 Cuadra listed four hundred. In 1990 Cuadra's list had expanded to nearly forty-five hundred (all these figures refer to databases in all fields of knowledge; the proportion of science/technology in 1990 was about 36 percent). The original type of text database—containing bibliographic references—was supplemented by other types, and Cuadra in 1990 was classing databases as follows:

- Bibliographic: references to documents (papers, reports, patents, dissertations, conference proceedings, books), which might be accompanied by an abstract
- Referral: references to organizations, individuals or nonprint media (e.g., audiovisual materials such as pictures, photographs, films, sound recordings)
- Numeric: data or statistically manipulated representations of data (e.g., time series)
- Textual-numeric: data that included both numeric and textual information (typical of handbooks)
- Full text: text of journal, newspaper, specification, encyclopedia articles, thesaurus, dictionary, etc.
- Software: computer programs

During the 1980s, bibliographical databases relating to special kinds of documentary material were produced—e.g., the World Patent Index,

Industry Standards, Engineering and Industry Software Directory, monographs in LCMARC and UKMARC, and databases listing conference papers, conference proceedings, and forthcoming meetings. Referral databases included those recording current research projects, such as MEDREP and ENREP. Full-text databases included Chemical Journals Online, the chemical encyclopedia Kirk Othmer Online, the pharmacopoeia Martindale Online, and the multilingual dictionary EURODICAUTOM. To locate the documents referred to in databases, online catalogues were developed, notably that at the U.S. Ohio College Library Center (OCLC), under the management of Frederick Kilgour.

As an example of the range of databases produced, in appendix D are noted 94 databases, from the 1990 Cuadra list, of particular relevance to environmental issues. In this sample, 38 were primarily bibliographical, 10 referral, 21 text/numeric or numeric, and 25 full text. The predominant producing country in the sample was the United States (especially for full text databases), and only the long-industrialized countries were represented. In the 1990 Cuadra list as a whole, the country distribution of database producers (excluding the Soviet Union, about which information was limited) was as in table 30, demonstrating the unevenness of database development.

In the 1980s, a new compact storage medium was developed, CD-ROM, upon which data could be far more densely packed than on magnetic disks. It became possible to fit large databases onto these disks and to have them accessed via software on microcomputer. The way was open (in principle) for such stores of information to be available for personal use on the scientist's desk. By 1990 Cuadra was listing four hundred commercially available CD-ROM publications, more than half of them corresponding to existing online databases. But new types of material were also represented: collections of sound or speech; full text that included figures, illustrations, charts, photographs and other graphic material; and collections of graphic

Table 30 Distribution of Database Producers, 1990

USA	1,124	Spain & Portugal	42
UK	174	Japan	37
France	152	Italy	30
Canada	108	Netherlands	27
Germany	105	Rest of Europe	41
Australia & NZ	73	Rest of world	22
Scandinavia	70		

Source: Cuadra (1990)

images without text. Some examples of CD-ROM material not available online at that time were: Languages of the World (full text dictionaries), Oxford Textbook of Medicine, and World Weather Data. The diskettes used in microcomputers also became a medium for the transfer of information—both software and data. In 1990, Cuadra listed sixty-six diskette products—some examples being Autofacts, Chemtox, and Perinatal Trials—and many others were also available.

The volume and variety of material available on CD-ROM and diskette increased greatly during the 1990s. Publishers realized that CD-ROMs were a means of providing easily searchable versions of bibliographies, dictionaries, encyclopedias, directories, statistical tables and other reference handbooks, and by 1996 there were hundreds of these available in the new format. Academic institutions, in particular, took up the CD-ROM technology, and through campus networks offered students and researchers a cheap way of accessing machine-readable information.

There was an explosion of material on and use of the Internet during the same period, especially prompted by much simplified access to resources via the World Wide Web and its associated software, which made it easier to contact an information source and to move from one source to another via hypertext links. Scientific publishers were now moving towards the provision of electronic versions of primary journals, including text, illustrations, chemical structures—in fact, everything that appears in the printed form. There were plans to create links between cited references in journal papers or bibliographies and the items cited. Scientific institutions of all kinds constructed "pages" on the Web giving information about their activities and facilities. Web sites were created through which subject searches could be made of what material might exist on a particular topic, using "search engine" software. An outline of the development of the Internet is given in appendix G.

SCIENTIFIC AND TECHNICAL DATA

As science and technology expanded throughout the twentieth century, and large national or international projects became common, the volume of data produced greatly increased, and the problems of access to such data became ever more pressing. In 1966, UNESCO and ICSU established an international Committee on Data for Science and Technology (CODATA), to improve the quality and accessibility of data and to facilitate international cooperation. One of its tasks was to identify the categories of data with which the committee was concerned: not only quantitative data (e.g., physical, chemical, meteorological, physiological) but also qualitative (such as chemical structures, classes of rocks, stellar spectra, biological taxonomy, gene sequences). The data could be in the form of numerical values, diagrams, graphs, models, maps or symbols. It could be primary observational or experimental data, or derived by theoretical calculation.

In some fields—e.g., chemistry, meteorology, geological survey, astronomy—the volume of data had become immense. The amount of data produced in scientific research was far greater than could be reported in the publication system, and even that which was published was inadequately tagged and indexed, making access difficult. Some specialized journals devoted to data publication were produced, e.g., *Atomic Data and Nuclear Data Tables* (1969), the *Journal of Chemical and Engineering Data* (1959), and the *Journal of Physical and Chemical Reference Data* (1972). Broad data compilations such as the *International Critical Tables* could no longer hope to cover the whole range of data, nor even the more specialist data handbooks that supplemented them.

The use of computers to acquire, massage, archive, access, retrieve, display and evaluate data in a variety of forms has continuously expanded in recent years. Many numerical and textual/numerical databases became publicly available online, as noted earlier, but at the same time many more were built for access by specialists only. For example, in 1977 the U.S. Geological Survey had over fifty systems handling geoscience data in digital form, totaling perhaps half a million "bits" (binary digits). It was estimated that by the late 1990s there would be recorded every second 10 billion bits of global-monitoring data. The number and variety of data sources that became available led CODATA to set up a World Data Referral Center to record them.

It is not easy to document the development of these restricted databases, but one example may be given: astronomical data. Collections of data about the positions and properties of stars have a long history. Clay tablets from Uruk in Mesopotamia, dated to 1650 B.C., provided observed positions of Venus. But detailed, documented observations date from the sixteenth-century Danish astronomer Tycho Brahe. In the following centuries, many star catalogues were published, a few of the notable contributors being John Flamsteed (1712), James Bradley (1762), Friedrich Bessel (1818) and Johann Dreyer (1888). By 1990, several thousand reliable catalogues were in existence, ranging from the massive series of *Durchmusterung* of Friedrich Argelander (1859–62), David Gill and Jacobus Kapteyn (1895–1900), and J.M. Thome (1892–1932), and the general *Smithsonian*

Catalogue of K.L. Haramundanis (1966), to the spectroscopic *Henry Draper Catalog* (Annie Cannon and S. Pickering, 1924), and on to many specialized lists for particular types of stellar object.

During the 1960s, there began the conversion of star catalogues into database form. In 1970, the International Astronomical Union set up a bureau to distribute information on machine-readable catalogues. Two years later, the Centre de Données Stellaires (CDS) was set up in Strasbourg and started to collect all such existing databases. During the next decade, similar data centers were set up in the United States, Germany, the Soviet Union, and Japan. The same stellar object was named in various ways in different catalogues, and a "cross-identification" file of synonyms had been started in 1971. The next step was to use this file to connect the different catalogues, and at Strasbourg an integrated database was established, SIMBAD, that combined all the information on over half a million stellar objects, derived from five hundred catalogues.

In addition to this database, others were compiled as a result of special astronomical projects, for example, the data collected by the International Ultraviolet Explorer (IUE) satellite (1978 on), which was housed in Spain; the archives of data from the EXOSAT (x-ray information) and Infrared (IRAS) satellites, launched in 1983, housed in the Netherlands; the Space Telescope archive in Germany; data archives at the U.S. National Space Science Data Center. Specialized communication networks for use by astronomers were developed during the 1980s, e.g., the British STARLINK, the Italian ASTRONET, the U.S. NASA Science Network, and the international Space Physics Analysis Network (SPAN). In 1988, the European Space Agency, based in Italy, announced a project to develop a European Space Information System (ESIS) giving astronomers user-friendly access to many of the astronomical databases, and to the bibliographic databases available on the ESA-IRS online host.

In 1993 the CDS made publicly available an astronomical information service on the World Wide Web (Egret and Albrecht). By 1997 it provided access to a variety of services: (a) a bibliography of half a million abstracts on astronomy and astrophysics; (b) Astroweb, a collection of pointers to over two thousand astronomically relevant Internet resources; (c) Star, a collection of directories, dictionaries, databases and related products on astronomy, space sciences and related fields; it included StarWorlds, six thousand entries for organizations, institutions, associations, companies; StarHeads, links to the Web pages of twenty-five hundred individuals; and StarBits, an extensive list of abbreviations, acronyms, contractions, symbols; (d) star catalogues—over fourteen hundred online, including SIMBAD, most also searchable via the ESIS tool for retrieving astronomical data.

STUDIES OF COMMUNICATION IN SCIENCE AND TECHNOLOGY

During the twentieth century—and more particularly, since 1950—scientists and those concerned with information provision in science began to study its communication structure in a systematic way. Better understanding of the structure provided guidance in the formulation of mechanisms to improve communication and information provision. In this section, a brief account will be given of some aspects of these studies.

In this book, the term "science" has been used broadly to include technology, agriculture and medicine, and consequently the term "scientist" might encompass the researcher, the engineer, the agricultural adviser, the physician and other social roles. One line of development in communication studies has been to analyze these roles in more detail, identifying different kinds of scientific, technical and professional activity, which might give rise to different information requirements and communication patterns.

The broadest distinction has been between science as "pure research," carried out to understand natural phenomena without regard to any possible practical utility; technical development or "applied science," designed to create a new or improved product or process with practical utility; and industrial, agricultural or medical "practice," the actual provision of consumer products and services. The second and third roles have been grouped together as "technology."

Derek Price in 1969 made a broad distinction between scientists and technologists. The former were motivated to publish and hence their characteristic product was the scientific paper. The content of science was embodied in papers, and its historical development could be traced in its literature. Technologists, on the other hand, were motivated to create products. The content of technology was embodied in those products, and in a sense it had to force itself into a written form: despite the existence of patents, technologists were often reluctant to publish accounts of their inventions. Although there were many technical journals, they had "a newspaper-like current awareness function . . . advertisements and catalogues of products were the main repositories of the state of the art for each technology." Price considered that the technical report was an attempt to force technology into the same pattern of literature accumulation as science.

Research, development and practice have obvious differences in aims, nature of work, outputs, and information needs. We will give two examples of the kinds of study made that have illustrated these differences. First, Richard Rosenbloom and Francis Wolek in 1967 studied scientists and engineers in large U.S. industrial corporations. They demonstrated the existence of a continuum of characteristics, the two extremes of which were "highly professionally oriented" (nearest to pure research) and "highly operationally oriented" (the nearest to industrial practice). The two extremes differed markedly in their communication activities—for example, the researchers relied for information much more on published scientific literature and on contacts with other researchers outside their own corporation, whereas the practitioners relied much more on personal contacts, particularly those within their own corporation. Our second example is Thomas Allen, who in 1966 studied a series of large scientific and engineering projects in the United States and reported that the scientists in his sample on average spent about 20 percent of their time with published literature and derived half of their ideas from it; the engineers spent only 8 percent of their time in this way and derived only 8 percent of their ideas from the literature.

Ronald Havelock in 1969 identified different kinds of information generated by the three social roles mentioned above. The pure researcher produced factual data (the physical properties of chemicals, the attributes of animals and plants, etc.), generalizations (theories, laws, principles), and scientific methods and procedures. Technical development produced principles to guide practitioners, engineering data, technical methods (such as systems analysis), designs for new products, and prototypes. The practitioner produced actual products (manufactured items) and services, and what Havelock called "software" (computer programs, but also instruction manuals, training programs, etc.). All these outputs could be sources of information to others.

Havelock further analyzed information flows within the research, development and practice worlds.

1. Researchers relied mainly on each other for information (via published papers or various forms of personal contact), though they had to turn to industrial practice for the instruments they used in their work; new scientific problems tended to arise from previous scientific results, but researchers could be stimulated by problems arising in development or practice.
2. Technical development received from pure research new knowledge that it could apply, and it learnt of current practical problems that

needed solution; with the growth of complex technology, development needed to draw on a variety of pure sciences and to combine a variety of technological procedures, so that information flow between the sciences and technology, and between different branches of technology, could also be complex.

3. Practitioners received aid and information from technical development, and feedback from consumers as to the effectiveness and efficiency of their products and services. John Holmstrom in 1954, writing specifically with reference to Britain, charted relationships between organizations concerned with the transfer of technical information (figure 10).

There have been studies of differences in communication behavior amongst scientists working in the same field. W. Hagstrom in 1965 developed a typology of scientific researchers as communicators, for example:

- scientific statesmen, with established reputations in their own fields, communicating at "policy" level with specialists in other fields
- highly involved leaders of their own fields, much occupied with travel, meetings and other forms of communication
- intradepartmentally oriented scientists, communicating mainly within their own departments
- productive isolates, alone in a research specialty, using the literature extensively
- student-oriented scientists, concentrating on teaching.

Thomas Allen in 1969 analyzed the technical-information contacts within research-and-development laboratories, and obtained new evidence for some of these categories. He identified four groups of scientific staff in a laboratory:

- "gatekeepers," who were much consulted by their colleagues ("communication stars"), and with many links to sources outside the organization ("cosmopolites")
- "liaisons," often consulted by colleagues, but mainly with links to other departments in the same organization
- the bulk of the laboratory staff
- isolates, with whom no one initiated contact.

Many studies have confirmed the continuing importance of personal, informal contacts within both science and technology. A survey of geoscientists and engineers by Alina Gralewska-Vickery in 1975 demonstrated

Figure 10 The Organization of Technical Science

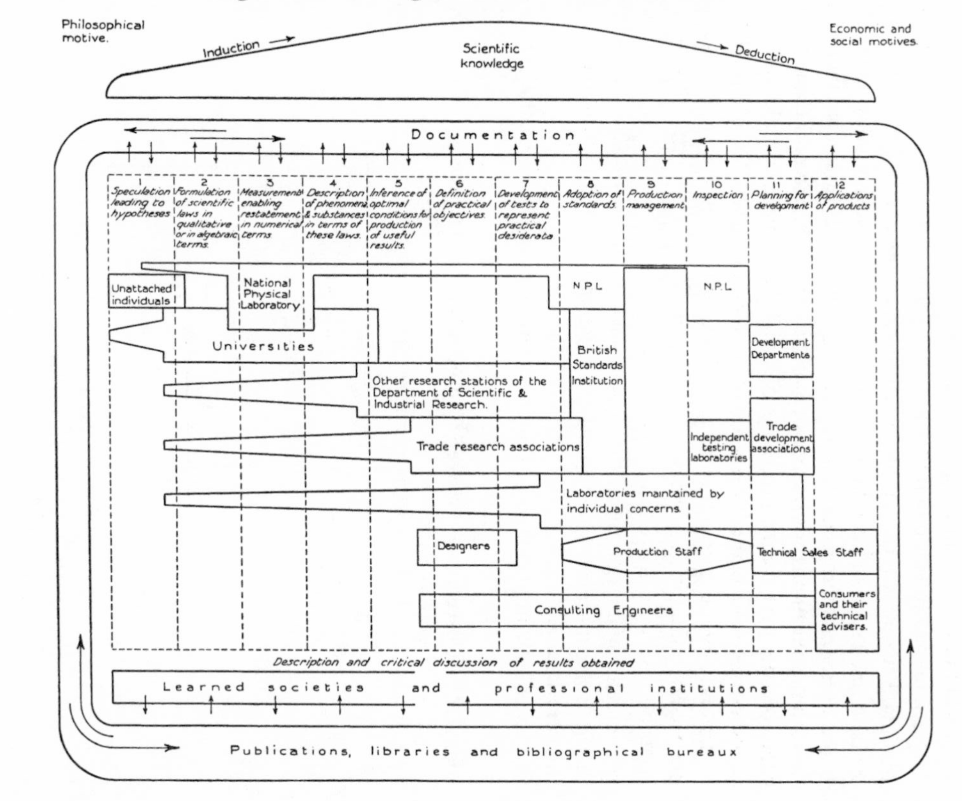

that personal contact was not at the expense of formal information activities such as consulting the literature: all forms of communication activity appeared to be correlated—a communication "star" used many forms (figure 11).

The same study examined the nonlocal personal communication of an individual scientist, a U.K. university professor: on average, this amounted to about one letter and one face-to-face contact per working day. One-fifth of the correspondence and one-third of the personal contacts concerned educational problems; one-tenth of both correspondence and contacts related to visits to or by the professor. Nearly half the contacts and one-sixth of the correspondence concerned advice and consultancy. Only a quarter of the correspondents and contacts were from the United Kingdom: in all, there were communication links to thirty-eight other countries, the major contributors being the United States, Canada and Australia. Half the correspondents and a quarter of the contacts were university-based; a quarter of the correspondents and nearly three-quarters of the contacts were industry-based.

The overall importance of communication in the work of the scientist

Figure 11 Correlations between Communication Activities

More opportunity to read
More reading
Less attempts to solve problems himself
More publishing
More society membership
More interest in research news
More use of abstracts
More journals used
More literature searches
More personal index of references
Less use of internal reports
More lecturing
More use of information centre
More report writing
More use of reviews
More use of trade literature
More use of external library
More use of bibliographies
More interest in organization news
More team discussion
More organizational library services
More use of books
More reprints file
More use of external reports
More work alone
More attend. meetings
More dissatisfied with journals
More visits
More interest in equipment

was indicated in a 1958 analysis by Russell Ackoff and Michael Halbert of chemists in U.S. industry: on average, each chemist spent over 40 percent of his time on work-related communication (compared to less than 40 percent on work at the bench and data treatment); of the strictly scientific or technical communication, over half was oral, 15 percent was writing, the rest reading. Within the averages, figures for individual chemists varied widely—the percentage of work time spent communicating ranged from 16 percent to over 70 percent.

Careful analysis of the process of scientific investigation was made in 1971 by Jerome Ravetz, from which a picture of the research process may be derived (figure 12).

In the course of an investigation, the research scientist seeks in the literature or informally to find hypotheses, methods, tools, and supporting information and theories, to help him establish a new fact or theory. William Garvey and his colleagues in the 1970s carried out studies relating the stage

Figure 12 The Research Process

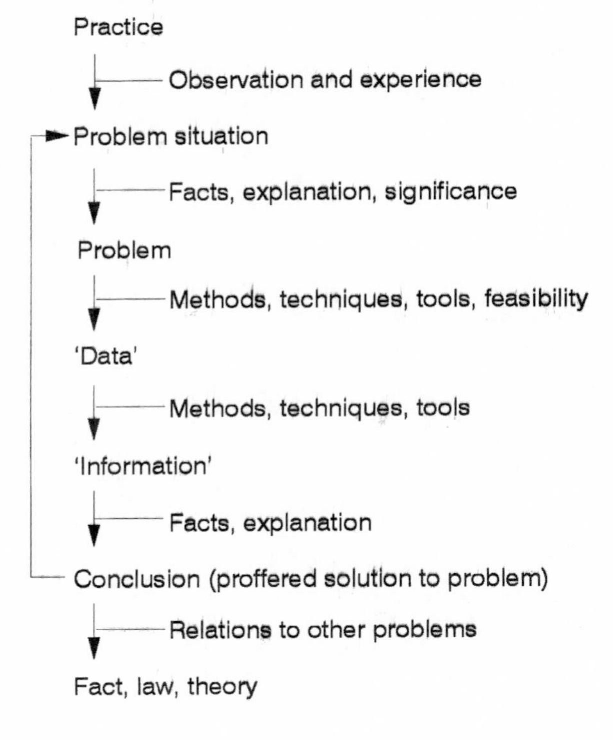

of a research process to the sources used by scientists to gain information. We may divide the sources into personal (contact with local or distant colleagues, attendance at meetings, preprints from colleagues), research reports (papers at meetings and in journals, technical reports), and consolidated science (books, etc.). Proportional uses of these sources were as in table 31.

All information sources were most used in problem identification, and later in linking the results into established knowledge. In general, personal sources were somewhat more favored than formal research reports, though the latter were heavily used for relating and integrating findings. Consolidated sources such as books had a minor but regular role throughout.

Ravetz underlined that the research process involved two different sorts of knowledge: the properties of the subject of enquiry, and the methods whereby the enquiry was conducted. Research reports embodying conclusions about the subject of enquiry were published in journals and would include some reference to methods used. But the detailed craft knowledge of the scientist could not be thus transmitted: for this, there was no substitute for interpersonal communication.

Studies on the communication of innovations summarized in 1971 by Rogers and Shoemaker showed that the same was true for the transfer of technological skills: competence was mainly accumulated in people, in tacit

Table 31 Research Stages and Sources Used

Figures show the percentages of respondents using source

Stage	*Sources*		
	Pers.	Res. rep.	Cons.
Problem perception/definition	71	70	22
Initial formulation of solution	45	49	16
Strategy for data collection	33	24	8
Choosing data-gathering technique	28	24	7
Designing equipment/apparatus	25	21	6
Choosing data-analysis technique	30	29	12
Relating data to known information	58	86	19
Interpreting collected data	53	48	14
Integrating findings into science	59	84	17

Note: Pers. = personal sources
Res. rep. = reports of research
Cons. = consolidated sources
Source: Garvey (1979)

and intangible form. The technical performance of firms and nations depended on skills that could often only be transferred by experience and demonstration. For example, in developed countries, knowledge of new technical innovations that could be applied to practice was transferred through documentary sources only 50 percent of the time—interpersonal contact was the source for the rest. In less developed countries, documentary sources accounted for only 30 percent of the transfers. When actual decisions to adopt an innovation were considered, personal contact with an adviser of some kind accounted for 80 percent of them in developed countries, and 90 percent in the less developed.

The cumulativeness of the scientific process has for a long time been made visible by the tradition among scientific authors of giving references in their books and papers to previous publications that have provided data, methods and ideas relevant to their own work. Analysis of these citations has thrown light on the structure of science. Sets of citations (e.g., all those appearing in the papers published over a period of time in a particular journal or set of journals, or those in a comprehensive treatise or monograph or bibliography) have been analyzed to provide statistical data. The creation of the *Science Citation Index* by Eugene Garfield in 1963, which brought together each year the citations occurring in many tens of thousands of journals, provided a valuable source of citation data for analysis. Some of the topics studied have been:

(1) *The age distribution of cited publications, giving an indication of how far back in time the citing authors have gone in tracing influences on their subject.* In very general terms, it has been found that the likelihood of a paper being cited decays exponentially as it grows older. This decay has been expressed quantitatively as a "half-life"—the time during which an initial citation rate drops to half the original value. Half-life figures varied from one subject field to another (Burton and Kebler, 1960). They were greater in mathematics, botany and geology than in physics and engineering, implying that relatively more use was made of older literature in the former group of subjects.

This general pattern has been modified by more-detailed analysis. A study by Derek Price in 1965 on "networks of scientific papers" showed that the steady decline in the rate of citation set in after an initial period of about fifteen years: before this, the rate was higher. Price interpreted this as the division of published literature into the "research front" (the last fifteen years or so of publications) that contributed actively to the continued growth of science, and a "classic" archive of all former publication into which the scientist dipped from time to time. A study in 1979 by Belver

Griffith and colleagues suggested that the drop in citation rate did not continue after twenty-five years: beyond that, the older literature was used fairly evenly, regardless of age.

(2) *The languages and/or countries of publication of cited items, giving an indication either of the actual international spread of a subject field or of the authors' recognition and acknowledgement of this spread.* Studies often showed that citation practice was too provincial—authors in one country might rarely cite those in another.

(3) *The subject areas of the cited items in relation to those of the citing items, giving an indication of interrelationships among subjects.* Earle and Vickery in 1969 made a broad distinction between "self-citation" (the degree to which publications in a subject field cited other publications in the same subject field) and "self-derivation" (the degree to which citations to items in a subject field came from items in that field). High self-citation indicated that a subject field was relatively independent of others in its development. The study provided data that suggested that some subject fields were relatively highly self-citing (e.g., botany 86 percent, mathematics 79 percent, physics 75 percent) whereas for others self-citation was lower (e.g., zoology 38 percent, geology 52 percent). Low self-derivation indicated that a subject field contributed much to the development of others. Some fields had relatively low self-derivation figures (e.g., mathematics 6 percent, chemistry 30 percent) whereas for others they were higher (e.g., astronomy 93 percent, geology 79 percent).

Herbert Small and Belver Griffith in 1974 used cocitation in published documents as a measure of linkage between two earlier items. On this basis they developed maps of links between subjects, at both general and specific levels. At a high level, for example, they demonstrated the close links between biomedicine and chemistry (cocitation strength 1002), or chemistry and crystallography (strength 613), compared to the weak links between, say, biomedicine and solid state physics (strength 3) or crystallography and geology (strength 11). At a lower level, they obtained bibliographic confirmation of the existence of specialty clusters within biomedicine (e.g., immunology, cancer, proteins, bacterial DNA) and the links between them.

THE CURRENT SCIENCE-INFORMATION SYSTEM

The cycle of information generation, processing and use in contemporary science has been summarized in figure 13, taken from the *Encyclopedia of*

Library and Information Science. The cycle represents the mechanism whereby scientific information is steadily cumulated and integrated. The inner circle suggests the timescale in which this is achieved.

At the top is the research investigation considered in an earlier section. During its progress, information may be recorded in laboratory notebooks and diaries, and it may be accompanied by informal correspondence and memoranda to supervisors or sponsors. When results have been achieved, they may be announced in a letter to the editor of a journal such as *Nature*, or a contribution to a letters journal. If the results have a potential commercial value, a patent claim may be lodged.

The first full announcement of an investigation may be made at a scientific conference, from which published preprints, proceedings and reprints may emerge; or it may appear in the form of a technical report or a dissertation or thesis; or—either alone or in addition to the other announcements—the results may be published as a journal article, perhaps accompanied once again by a preprint and/or reprint.

"Surrogation" refers to the appearance of announcements of the existence of the article, report, dissertation, proceedings paper or patent—in a bibliography, an abstracting/indexing service, or a "current awareness" service (SDI), all of which may be in printed and/or electronic form. In due course, information from the research findings may be incorporated into

Figure 13 The Current System of Scientific Communication

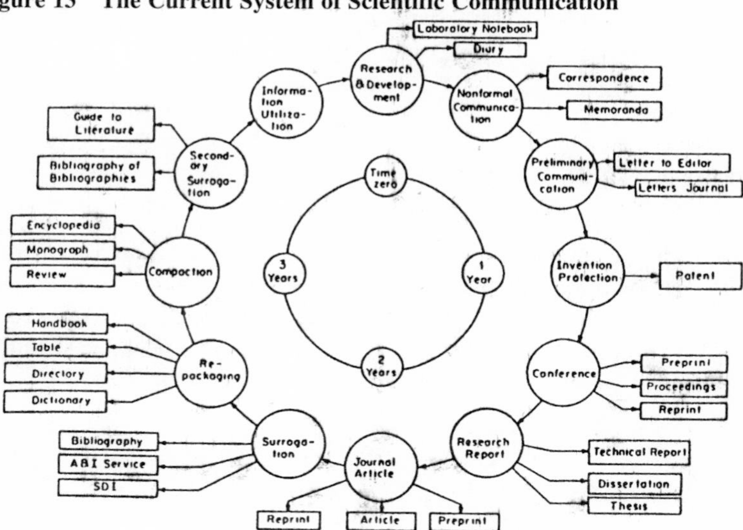

Figure 14 Formal Information Channels

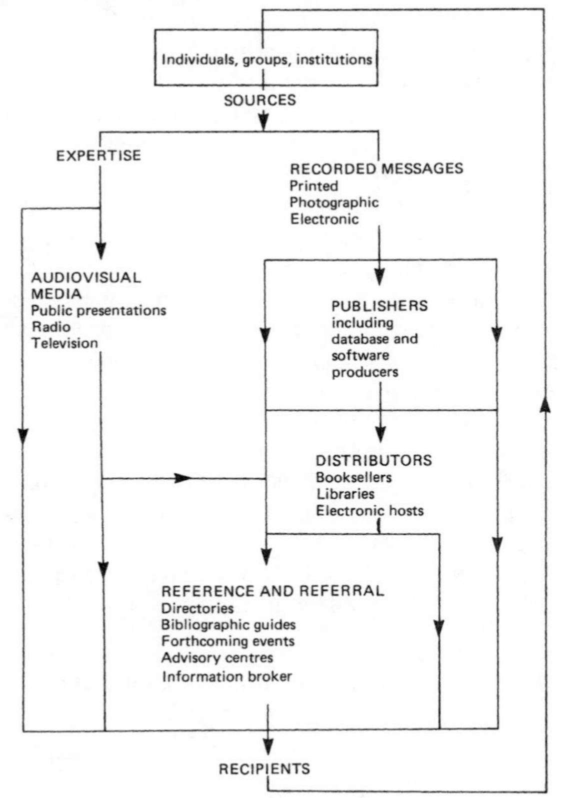

some compendium such as a handbook, data table, monograph or encyclopedia, or be discussed in a review paper. Any of these secondary "surrogates" may be listed in a bibliography of bibliographies or a guide to the literature. Tracing back through the cycle, a new researcher may come across the original results and use the information in a new investigation. The information channels used may be summarized as in figure 14.

Section 8

Retrospect and Reflections

RETROSPECT

In very summary fashion, let us look back at the development of scientific communication. The history will be divided into seven periods: the earlier civilizations, classical culture, the medieval period, what may be called the scientific revolution (1450–1700), the eighteenth, the nineteenth, and the twentieth centuries. Within each period will be very briefly noted: (1) some historical background, (2) major scientific and technical activities, (3) social roles of relevance to science and its communication, (4) communication mechanisms that developed during the period, and (5) advances in communications technology.

The Earlier Civilizations (to about 600 B.C.)

The early human settlements grew into cities, and thence into empires in Mesopotamia, Egypt and elsewhere.

The technology developed up to this point included agriculture, animal husbandry and fishing. Irrigation to water the cultivated fields was well advanced. Crafts such as spinning, weaving, pottery and carpentry had developed. Metals were mined and worked. Buildings were constructed of brick and stone, including pyramids and temples. In science, arithmetic and geometry started to develop; there was early interest in astronomy and the calendar. Medicine was practiced.

Social roles included craftsmen, architects, physicians, and scribes. They communicated by personal contact, and perhaps used messengers. Archive collections were built up, some including technical texts.

Communications technology was confined to the art of writing on such media as clay tablets, papyrus and stone, and aids to travel: roads, wheeled vehicles and sailing ships.

Classical Culture (600 B.C. to A.D. 500)

New city-states developed (Phoenician, Greek), and new empires (Persian, Roman). There was much movement of peoples into and around Europe. Christianity spread as Rome and the economy of Europe declined.

A scientific outlook was established. Mathematics and mechanics were strongly developed, astronomical systems were constructed, descriptive biology and geography flourished, rational medicine was practiced. Both architecture and engineering advanced.

Relevant social roles included natural philosophers, some full-time professional scientists, engineers, and physicians, and commentators and encyclopedists. Publishers and booksellers came on the scene.

New communication mechanisms appeared: research and teaching academies, libraries, bookshops, and the scholarly editing of texts. Developments in communications technology included the alphabet; the first use of parchment; improved road networks, carriage and ship transport; maps and charts; and plant illustrations.

The Medieval Period (500–1450)

Western Europe, Byzantium and Islam at first developed separately. The European economy revived, with the growth of new towns and the development of new crafts. The three previously separate areas then interacted.

Science and technology in many ways marked time, but there were advances in algebra and optics, and in Arab medicine. Alchemy flourished. Cathedrals, castles, palaces and mosques were built. Water wheels and windmills came into use. Mechanical clocks were developed.

Social roles included Christian scholars, Arab philosophers, medical doctors, university teachers and encyclopedists, and craftsmen such as masons.

New communication mechanisms were monastic scriptoria, universities and their book trade, academic libraries, and considerable translation activity. The vernacular languages of Islam and Europe became literary vehicles. Technical developments in communication included the parchment codex, the use of paper, and block books.

The Scientific Revolution (1450–1700)

Capitalism began to replace feudalism in Europe. Nation-states developed. Europeans explored the world and established colonial empires.

Experimentalism established itself in science. The mathematical calculus was developed, there were advances in understanding the solar system and gravitation, in physics, anatomy, microscopy and natural history. Scientific instruments were constructed. Chemistry and the beginnings of chemical manufacture appeared. Engineering and mining further developed. Navigation advanced, and the construction of maps.

Social roles included amateur scientists (and some at universities), intelligencers and journal editors, bibliophiles and bibliographers, printers, publishers and booksellers, instrument makers, explorers.

New communication mechanisms were extensive correspondence, scientific societies and their meetings, the first journals and bibliographies, travel and exploration, zoological and botanical gardens. Patents were established. The outstanding new communication technologies were printing and the woodcut, engraving and etching.

The Eighteenth Century

Capitalist manufacture developed. Agriculture and industry were revolutionized by technical advances, beginning in Britain, then in Europe. The United States broke free from Britain and began to develop. International trade became important.

Science began to consolidate itself: modern chemistry and biological classification were established and electricity was discovered. Steam power was introduced, textile machinery and metallurgy developed. Popular interest in science appeared.

The social roles of university and amateur scientists continued, scientific publishers appeared, mechanics and artisans flourished, industrial entrepreneurs became important.

Journals, bibliographies and societies continued to increase. Technical encyclopedias developed. Museums were created. River and sea transport increased.

The Nineteenth Century

Modern industry began to develop, particularly in Britain, Germany and America. World trade was dominated by the industrial nations.

In technology, factories came on the scene, and transport was revolutionized. The atomic theory, organic chemistry and the manufacture of dyestuffs and explosives developed. Energy, electromagnetism and thermodynamics were explored. Geology and evolution theory developed. Physiology was studied, and the germ theory of disease was established.

Science and engineering became more professional. Links between science and industry were developed. The profession of librarian was established.

Communication mechanisms included abstracting and indexing serials, reviews of progress and the expansion of patents. New forms of technical institution developed. Technical skills were transferred from one country and continent to another.

Advances in communications technology included the railway and steamship, the telegraph, telephone, typewriter, printing mechanization such as power presses, papermaking machines, the linotype and monotype, and illustration processes—lithography, photography, photoengraving.

The Twentieth Century

Large-scale and multinational industry developed. All countries were drawn into the world economic system. There was increasing disparity in levels of economic development. Major wars and revolutions occurred. Former colonial countries became politically independent. Socialist states were set up, now (temporarily?) in decline.

A few highlights of science have been the quantum theory, relativity, the expanding universe, microbiology, molecular biology and genetics. Radioactivity was followed by nuclear energy, electronics by computers and automation. Polymers and plastics were developed, as were automobiles, aircraft, and then space exploration. Science began to spread worldwide, interacting strongly with technology.

Social roles included academic researchers, industrial and government scientists, development and consultant engineers, agricultural advisers, information scientists and database publishers, computer host operators. Specialization has grown apace.

Communication mechanisms have been industrial and government research laboratories, conferences of all kinds, newsletters and letters journals, technical reports, data compendia, electronic databases. Formal international links have developed. Libraries were supplemented by information centers and data centers.

New communications technology has included the cine film, radio and television, office duplicating, photocopying, offset lithography, microforms, photocomposition, facsimile, interactive computers, computer typesetting, word processors, desktop publication, telecommunication networks, satellites, electronic mail, magnetic and optical data storage, computer graphics.

The study of communication has led to the creation of a new discipline, information science, with links to sociology, psychology, linguistics, the cognitive sciences generally, and to computer science.

REFLECTIONS

Science emerged from the speculations of the "wise man" and the skills of the craftsman. It combined theory with practice—hypothesis and generalization with observation and experiment. In time, scientific thinking freed itself from religious and philosophical constraints, and developed autonomously. Technology for a long time also developed independently, little influenced by its contemporary science. Though science from time to time took note of practical problems, it was not until the Renaissance that scientists began to focus on the contribution they could make to technology, and not until the nineteenth century that effective contributions were made. In the present century, science and technology interact intimately. Every aspect of life—industrial, agricultural, medical, even administrative—increasingly needs the help of organized science for its day-to-day operations, and even more for its progressive development.

Science has grown over the centuries into a social institution with its own tradition and discipline, its own specialized workers, and its own sources of funds. Since scientific research in itself does not generate income, patronage of some kind has always been necessary. But through much of its history, the control of science has been in the hands of scientists. In some ways science has now begun to lose its independence. Because much research now proceeds within industrial and government establishments, and because research often involves heavy expenditure, the scientific process has itself become industrialized. Many scientists are now employees, working on a project for the benefit of their employers; or individuals existing on a succession of grants from funding agencies that determine what research is worth funding. The old tradition of free scientific enquiry is compromised, and some see this as inimical to the future health of science.

Communication mechanisms in science, such as journals, have also in the past largely been under scientific control. But information provision has now become as subject to industrialization as science itself. Publishing and printing have, of course, always been commercial activities, but the control by societies and academia of scientific publications and of their distribution has in the past infused these activities with something of the communal ethos of science. The increasing control of information provision by large agencies oriented towards "the market" could lead to less than adequate provision to areas of science and technology with slender financial resources.

As scientific activity has grown in volume, it has inevitably become subdivided into specialties, as has been illustrated at some length in earlier sections. Yet at the same time, the increasing interpenetration of science and technology has begun to enforce the need for multidisciplinarity. Technical and practical problems can rarely be solved using knowledge derived from a single scientific specialty—the engineer typically has to combine elements from a number of fields to construct a product or system. Moreover, science itself has begun to tackle problems that require a multidisciplinary approach—for example, the study of the oceans, which has physical, chemical, biological and geological aspects. These developments have created the need for scientists from different specialties to work together in teams, and the further need for scientific "generalists" who can combine the contributions of the specialists.

Scientific knowledge and activity has grown exponentially—that is to say, every twenty years or so, it has doubled in size. It has become ever more difficult to impart in a university course an adequate understanding of a scientific discipline at one moment in time, and the problem for the individual scientist of keeping up to date in anything outside his own specialty has become very great—at the very time, as just seen, when an understanding of relations between different fields has become more important.

The growth of specialization in work has been accompanied by the emergence of societies, research institutions, journals, monographs and compendia ever narrower in scope, to cater for the specialist. But the scientist may have to go outside his own specialty when seeking information on methods and tools of data collection and analysis, and in the interpretation of results. Consequently there has been a need for the more general sources of information to be maintained. An illustration of this is *Chemical*

Abstracts, which since its origin in 1907 has endeavored to cover the whole discipline of chemistry, but since 1976 has also produced a series of *CA Selects*, each devoted to a specialty.

Since journals exist at every level of generality, each of which may contain papers on specific topics, scientists have become increasingly aware of the problem of bibliographic scatter. In every specialty, a relatively few "core" journals are devoted to the subject, but other relevant papers are scattered throughout a much wider range of journals. The mathematics of the distribution was first investigated by S.C. Bradford in 1934, and has since been much explored. The existence of the phenomenon has been copiously documented.

The specialization of science initially consisted in the successive subdivision of the established disciplines. But interdisciplinary subjects developed—e.g., biophysics, radiochemistry—giving rise to new specialties, with their own specialist societies and publications. This accentuated bibliographic scatter, since biophysics papers could be, and were, published in either biology or physics journals; and it complicated bibliographic access, since such papers could be, and were, announced in either biology- or physics-abstracting serials.

In the 1960s, a contrast was made between mission-oriented areas of scientific and technical activity (such as nuclear energy or space exploration) and the well-established disciplinary areas (physics, chemistry, etc.). A "mission," in this sense, was a large-scale technology project, drawing on a range of different scientific disciplines and engineering skills. For some such missions, information systems were built up—journals, compendia, abstracting serials, databases—and in fact we have seen earlier that two of the first databases available online were the abstracts of NASA reports and of nuclear science and technology. These developments further contributed to scatter and to duplication of the announcement of scientific and technical results.

More recently, multidisciplinary subject areas and technological "missions" have increased in number, at a rate that has not been reflected in the publication pattern. To keep abreast of the published information in such a subject area, it has become necessary to search primary and secondary publications in several scientific disciplines and fields of technology. An example of the problem occurred during the study of cosmic background radiation: teams at Princeton University and Bell Laboratories developed microwave apparatus to detect such radiation and announced their success in 1965 in an issue of *Astrophysical Journal Letters*. They made no mention of the theoretical predictions of the radiation by George Gamow and his colleagues, published during the 1940s in *Physical Review* and elsewhere. They later explained: "We read only the microwave journals, so we never saw any of Gamow's stuff; there was no cosmology literature, so scientific papers on the problem were published all over the place. We simply did not do our homework."

The difficulties of multidisciplinary search led the International Council on Scientific and Technical Information (an ICSU committee) in 1989 to set up a study group on the interdisciplinary searching of databases (Weissgerber). One problem area that they have identified relates to the terminologies used in the different subject fields and to the index terms used in the different databases for search purposes. For example, there is still lack of commonly accepted and applied nomenclature and terminology in biological, chemical, geological, geographical and engineering fields; terminology in rapidly developing fields such as biotechnology or materials science often evolves in meaning and specificity, thus complicating the task of searching over a stretch of time. These difficulties serve to aggravate the problem faced by all interdisciplinary search, that each scientific and technical specialty develops its own jargon with which outsiders are unfamiliar. The sheer size of the vocabulary of science and technology has become daunting, and the variety of languages in which science is now recorded compounds the problem.

In the world today there are about 6 million research-and-development scientists, and over 130 million scientists and technicians in all, unevenly distributed over the nations, as we have seen earlier. In 1971 the UNISIST program of UNESCO, reporting on the feasibility of a world science-information system, set out a noble aim: "Science has come to dominate the life of all human beings: for this reason alone, the material on which it feeds—information, data—is a resource of vital importance to world society. . . . No individual, no class, no community should be barred from access to the stores of knowledge accumulated by mankind."

Scientists, technologists, practitioners, and the consumers of the products of science and technology all generate information of one kind or another and seek information produced by others. Between those who generate information and those who need it there now stands a complex network of communication channels. The network cannot truly be called an "information system": it is a loosely linked assembly of independent institutions—societies that produce journals and abstracts and hold conferences; commercial publishers and booksellers; government-reports centers; patent offices; database producers and online hosts; research establishments with in-house information stores; academic, industrial and government libraries, information centers and data centers; translation agencies; referral services; telecommunications services. It is true in principle that successful navigation through this network can ensure that "the whole of knowledge is available for everyone to use," as John Holmstrom said in 1956. But there are many barriers to use. "That communism of the intellect that is the beneficent glory of science" (Holmstrom's words again) meets many obstacles.

The multiplicity, independence and often the incompatibility of information services means that "planning" the communication system of science and technology has often involved no more than stating ideal goals (such as those of the UNISIST declaration) and suggesting guidelines and standards that institutions should ideally follow. Reality has all too often remained obstinately chaotic. In 1948, the convenors of the Royal Society Scientific Information Conference in London expressed the view that "the task of keeping up with scientific literature is becoming an impossible one, and is in turn leading to inefficiency and to a certain amount of frustration in scientific research and in the application of science." It may be doubted whether the situation has improved.

Some believe that modern technology offers a solution to communication difficulties. The 1980 UNESCO report, *Many Voices, One World*, prepared under the chairmanship of Sean MacBride, examined communication in general, and stressed that the new electronic technologies potentially opened a path for a new era in communication:

> Distance has ceased to be an obstacle, and the possibility exists of a universal communication system linking any point on the planet with any other. . . . It is feasible to envisage a web of communication networks, integrating autonomous, or semi-autonomous, decentralized units. The content of messages could be diversified, localized and individualized to a large extent. . . . The global web of electronic networks can, potentially, perform a function analogous to that of the nervous system, linking millions of individual brains into an enormous collective intelligence.

This harks back to developments envisaged in the 1960s. At the Massachusetts Institute of Technology, when work on time-sharing began (project MAC), an experiment in interactive information access (INTREX) was also started, led by Carl Overhage. The concept was developed of an "online intellectual community." All the members of a university community, including scientists and technologists in associated industry, could work in close and frequent interaction with a computer-based information system. The system would be at once a store, a processor and a transmitter of information—the central nervous system of the academic community, but itself only a node in a larger network serving a whole complex of institutions carrying out intellectual tasks. At the computer consoles on their desks, members of the community would think, study, teach, learn, calculate, analyze data, invent, solve problems, make decisions, carry out many other processes of cognition and computation, seek and retrieve information, prepare reports, publish results, interact with colleagues far and near.

Since that time, all the elements that could contribute to this concept have been developed: interactive computing of all kinds, computer-aided teaching and learning, innumerable programs of data manipulation, online search systems and databases, electronic document delivery, word processing, desktop and electronic publishing, electronic mail, bulletin boards, computer conferencing, local- and wide-area networks. Work has been carried out on the design of "work stations" at which such facilities could be provided. Certainly, these new information facilities have become essential to the business activities of multinational firms (Langdale, 1989). But for science and technology, is the online intellectual community near?

We must first note one general feature of the development of scientific communication: communication mechanisms, once established, rarely disappear. Personal contact was of necessity the first form of communication: its continued importance in science and technology has been clearly demonstrated. Meetings, visits and conferences, and the travel they entail, are as relevant as ever. Correspondence may have particularly flourished in the time of the intelligencers, but for leading scientists and engineers it remains a vital form of personal contact. Publishing, bookselling and libraries continue to be necessities. The translation industry flourishes as never before. Scientific societies, journals and serial bibliographies continue to proliferate.

The technology used to realize a social mechanism may change, but the old technology usually continues in a modified role. Written and typed letters and memoranda still exist, despite the word processor. Facsimile and electronic mail are new forms of correspondence, but the letter post remains heavily used in science and technology. We are not yet in a paperless society—the printing press is ever busy, and computer typesetting and desktop publishing only add to the torrent of paper. The drawing and the photograph serve to feed computer graphics. The telephone line is still the foundation of telecommunications.

Even if, within certain communities well endowed with high technology facilities, online interaction becomes of major importance, it is most unlikely to oust more-traditional forms of communication. For some purposes, these will remain valid and preferred. In particular, to gain access to information, the less well endowed majority will depend on the continued availability of publications in more traditional forms, and will continue to produce publications in these forms.

The development of electronic communications could proceed harmoniously. But *Many Voices, One World* pointed to factors that could distort and hinder communication. The new technology was, for the time being, "at the disposal of only a few countries and of a few people within them. The countries in which these discoveries originated still enjoy a massive advantage over other countries, where the development is proceeding in tortuous ways, hampered by a poverty which enforces a lack of the necessary infrastructure." These disparities could have serious consequences, the report maintained, in science and technology. The fact that most scientific and technical information was generated in and distributed by the industrialized countries could lead to a lack of availability of information needed for solving the problems of the less-developed countries.

Scientific and technical communication, as we have seen, needs to take many forms—electronics, print, writing, graphical illustration, speech, visual observation. Whatever the form, difficulties and inequalities exist—uneven access to electronic technology, delays in the transmission of printed materials, lack of travel resources, language difficulties. Throughout science, as in society in general, there are the favored few and the less-favored majority. No individual, no class, no community is actually barred from access to mankind's store of knowledge, but between aspiration and achievement there is still a large gap.

* * *

Four thousand years ago, the Egyptian scribe Ahmose began to pen what we call today the *Rhind Mathematical Papyrus*, in which, he wrote, would be found "rules for inquiring into nature, and for knowing all that exists, every mystery, every secret." Last year, each of several million research-and-development scientists sat down with a word processor, typewriter or ball-point pen to shed a little light on one of the myriad mysteries of nature. And in the epochs between?

> Yesterday, all the past. The language of size
> Spreading to China along the trade-routes; the diffusion
>   Of the counting frame and the cromlech;
> Yesterday the shadow-reckoning in the sunny climates.
> Yesterday the assessment of insurance by cards,

The divination of water; yesterday the invention
  Of cartwheels and clocks, the taming of
Horses. Yesterday the bustling world of the navigators. . . .
Yesterday the installation of dynamos and turbines,
The construction of railways in the colonial desert;
Yesterday the classic lecture
On the origin of Mankind. . . .
Tomorrow, perhaps, the future. . . .

—W. H. Auden

Appendix A

Universities

The formation of universities provides one picture of the spread of scholarship across the world. The existence of a university does not, of course, necessarily imply that it initially included teaching and research in science. The list runs from the ninth to the nineteenth century. Foundations in the twentieth century are too numerous to list here.

Ninth century
Salerno

Twelfth century
Bologna, Paris, Oxford, Modena, Montpellier

Thirteenth century
Perugia, Vicenza, Cambridge, Arezzo, Salamanca, Padua, Naples, Toulouse, Siena, Sorbonne, Lisbon, Macerata

Fourteenth century
Rome, Orleans, Florence, Grenoble, Pisa, Vallodolid, Prague, Pavia, Cracow, Vienna, Pecs, Heidelberg, Cologne, Ferrara, Erfurt

Fifteenth century
Turin, Leipzig, Marseilles, St. Andrews, Rostock, Besancon, Louvain, Caen, Poitiers, Rennes, Bordeaux, Palermo, Barcelona, Glasgow, Greifswald, Freiburg, Basle, Nantes, Munich, Genoa, Mainz, Uppsala, Copenhagen, Aberdeen, Tubingen, Catania, Santiago de Compostela, Zaragoza

Sixteenth century

Valencia, Wittenberg, Seville, Frankfurt, Madrid, Granada, Marburg, College de France, Coimbra, Lausanne, Nimes, Strasbourg, Verona, Konigsberg, Messina, Dillingen, Jena, Geneva, Lille, Sardinia, Helmstadt, Nancy, Berne, Leyden, Vilna, Edinburgh, Wurzburg, Graz, Dublin, Valletta, Urbino, Ljubljana, Malta

Americas: Santo Domingo, Lima, Mexico City, Bogota

Seventeenth century

Giessen, Oviedo, Groningen, Strasbourg, Salzburg, Amsterdam, Budapest, Utrecht, Helsinki, Ancona, Bamberg, Duisberg, Limoges, Durham (dissolved 1660), Lvov, Kiel, Lund, Zagreb, Urbino, Innsbruck, Halle, Cagliari

Americas: Manila, Cordoba (Argentina), Sucre, Harvard, Guatemala, Ayacucho, Quebec, Cuzco, William and Mary

Eighteenth century

Venice, Breslau, Mantua, Lyons, Dijon, Bourgogne, Gottingen, Stockholm, Erlangen, Mannheim, Moscow, Brussels, Palermo, Ecole Polytechnique, Royal Technical College (Glasgow), Braunschweig, Clausthal, Hannover, Stuttgart, Karlsruhe

Americas: Yale, Caracas, Havana, Santiago (Chile), Philadelphia, Princeton, Columbia, New York, Brown, Quito, North Carolina, Guadalajara (Mexico), Halifax (Nova Scotia), New Brunswick, Dartmouth College, Rutgers, Georgia, Louisville, Pittsburgh, Knoxville, Vermont

Nineteenth century

Europe: Vienna Technical University, Gent, Liege, Sofia, Lyngby, Clermont-Ferrand, Aachen, Berlin Technical University, Darmstadt, Athens, Trieste, Oslo, Warsaw, Bucharest, Goteborg, Zürich, Fribourg, Kharkov, Kiev, Tomsk, Aston, Bath, Belfast, Birmingham, Bristol, Dundee, Leeds, Liverpool, London, Manchester, Nottingham, Reading, Sheffield, Wales, Belgrade

Americas: Buenos Aires, La Plata, Santa Cruz, La Paz, Recife, Rio de Janeiro, Belo Horizonte, Montreal, Hamilton, Ottawa, Kingston (Ontario), Toronto, Western Ontario, Winnipeg, Medellin, Cartagena, Cuenca, San Salvador, Honduras, Nicaragua, Asuncion, Montevideo, Washington, Atlanta, Auburn, Boston, Bryn Mawr, California Institute of Technology, San Diego, Cleveland, Colorado, Cornell, Drexel, Tallahassee, Chicago, Indiana, Iowa, Baltimore, Bethlehem, Baton Rouge,

Massachusetts Institute of Technology, Miami, Evanston, Columbus, Pennsylvania, Purdue, Renselaar Polytechnic Institute, Seattle, Syracuse, Tufts, Berkeley, Boulder, Farmington, Amherst, Ann Arbor, Minneapolis, other U.S. universities

Asia: Beijing, Xian, Hangzhou, Allahabad, Bangalore, Bombay, Calcutta, Madras, Tokyo, Ishikawa, Kyoto, Beirut American University, Lahore, Istanbul

Australasia: Adelaide, Melbourne, Sydney, Tasmania, Melbourne Institute of Technology, Auckland, Canterbury, Dunedin, Wellington

Africa: Sierra Leone, Cape Town, Bloemfontein, Grahamstown, Pretoria, Stellenbosch, Johannesburg

Appendix B

Institutions and Journals in Nineteenth-Century Britain

Journal titles appear in italics.

1800s

Proceedings of the Royal Society of London
Royal College of Surgeons of England
Royal Horticultural Society (and its *Proceedings*)
Royal Society of Medicine
Royal Medical and Chirurgical Society
Retrospect of Philosophical, Mechanical, Chemical and Agricultural Discoveries
Geological Society of London (and its *Transactions*)

1810s

Annals of Philosophy
Botanical Register
Journal of Science
Institution of Civil Engineers (and its *Proceedings*)
Edinburgh Philosophical Journal
Cambridge Philosophical Society (and its *Proceedings*)
Charing Cross Hospital (and medical school)
Regent's Park College, London

1820s

Lancet
Royal Astronomical Society (and its *Memoirs*)
London Journal of Arts and Sciences

Technical Repository
Mechanics' Magazine
Register of Arts and Sciences
The Botanic Garden
Edinburgh Journal of Science
Gardener's Magazine
Zoological Society (and its *Transactions*)
Arcana of Science
Magazine of Natural History
Quarterly Journal of Agriculture
London Mechanics' Institute (and in many other towns)
Edinburgh School of Arts
Medico-Botanical Society
University College London
Kings College London
Belfast Natural History Society
St. David's College, Lampeter

1830s

Annals of Electricity
Mining Journal
Hooker's *Icones plantarum*
Annals and Magazine of Natural History
Botanical Miscellany
Entomological magazine
Royal Entomological Society (and its *Transactions*)
Royal Statistical Society (and its *Journal*)
Paxton's *Magazine of Botany*
Analyst
British and Foreign Medical Review
Civil Engineer
Magazine of Zoology
Cambridge Mathematical Journal
Polytechnic Journal
Royal Microscopical Society (and its *Transactions*)
Royal Geographical Society
Harveian Society
British Association for the Advancement of
Science Manchester Statistical Society
Ministry of Health

Institute of Building
Royal Institute of British Architects (and its *Journal*)
Manchester Medical Society
Botanical Society
Yorkshire Geological Society
Royal Agricultural Society
British Medical Association
University of Durham
Geological Survey
Geological Museum
Westminster Hospital (and medical school)
Middlesex Hospital (and medical school)

1840s

Cambridge and Dublin Mathematical Journal
Pharmaceutical Journal
Retrospect of Practical Medicine and Surgery
Chemical Society (and its *Journal*)
Artisan
Chemical Gazette
Zoologist
Phytologist
Institution of Mechanical Engineers (and its *Proceedings*)
Florist
Practical Mechanics' Journal
Journal of Gas Lighting
Military College of Science
Huddersfield Technical College
Pharmaceutical Society
Ray Society
Royal Engineers Corps
Queens University Belfast
School of Pharmacy London
Royal College of Veterinary Surgeons
Royal Agricultural College
Architectural Association
Paleontographical Society
Bedford College London
Royal Botanic Gardens, Kew
Medical News and Library

Annual Report of the Progress of Chemistry
Annual of Scientific Discovery

1850s

British Medical Journal
Pathological Society (and its *Transactions*)
Chemist and Druggist
Engineer
Ibis
Liverpool Photographic Journal
Photographic News
Quarterly Journal of Microscopical Science
Photographic Journal
Daguerrian Journal
Naturalist
Society of Engineers
Geologists' Association (and its *Proceedings*)
Geological Survey and Museum
Recreative Science
Owen's College, Manchester
Royal Meteorological Society
Royal Photographic Society
Victoria and Albert Museum
Royal Scottish Forestry Society
Patent Office
Royal College of Science
Institution of Engineers and Shipbuilders, Glasgow South Wales Institute of Engineers
University College Newcastle
Merchant Venturers' Technical College, Bristol
St. Mary's Hospital (and medical school)
Birmingham and Midland Institute
Bath Technical College
Society of Public Health
London Natural History Society
British Ornithologists Union
British Horological Institute
Science Museum, London
Camborne School of Mines
Meteorological Office
Public Libraries Act

1860s

Chemical News
Geological Magazine
Journal of Anatomy
Nature
Colliery Guardian
Engineering
Entomologist
Record of Zoological Literature
Meteorological Magazine
Practitioner
Journal of Botany
Scottish Institution of Shipbuilders and Engineers
Institution of Naval Architects
Messenger of Mathematics
Popular Science Review
Institute of Gas Engineers (and its *Transactions*)
Quarterly Journal of Science
London Mathematical Society (and its *Proceedings*)
Surveyors' Institution
Iron and Steel Institute (and its *Journal*)
Hartley Institution, Southampton
Quekett Microscopical Club
Royal Aeronautical Society
Royal Commonwealth Society
Natural History Museum, Nottingham
Index to Foreign Scientific Periodicals in the Patent Office Library
Royal Army Medical College

1870s

Analyst
Chemical Review
Electrician
Metal Worker
British Veterinary Journal
Mineralogical Magazine
Journal of Physiology
Observatory
Institution of Electrical Engineers (and its *Journal*)
Medical Times and Gazette

Institute of Municipal Engineers
Physical Society (and its *Proceedings*)
Geological Record
Brain
Poultry World
Printing World
Contract Journal
Mathematical Association
University College of Wales
Yorkshire College of Science
Royal Society of Health
Portsmouth School of Science and Art
Society for Analytical Chemistry
Physiological Society
Mineralogical Society
Royal Institute of Chemistry
Annual Record of Science and Industry
London Medical Record
Library Association
Regent Street Polytechnic

1880s

Institute of Marine Engineers (and its *Transactions*)
Annals of Mathematics
Marine Biological Association (and its *Journal*)
Annals of Botany
Journal of Comparative Pathology
Institution of Mining Engineers (and its *Transactions*)
Kew Bulletin
City and Guilds Engineering College
Woolwich Polytechnic
Society of Chemical Industry (and its *Journal*)
British Dental Journal
Journal of Laryngology and Otology
Veterinary Record
British Journal of Dermatology
British Veterinary Association
Royal Forestry Society
Edinburgh Mathematical Society
North-East Coast Institution of Engineers and Shipbuilders

Society of Dyers and Colourists
Institute of Brewing
Anatomical Society
Ministry of Agriculture and Fisheries
University College Liverpool
University College Dundee
Mason Science College, Birmingham
Ophthalmological Society
British Dental Association
Institute of Patent Agents
Scottish College of Textiles
Royal Institute of Public Health
Royal Society for Protection of Birds

1890s

Institute of Mining and Metallurgy (and its *Transactions*)
Institute of Sanitary Engineers
Institute of Water Engineers
Institute of Heating and Ventilating Engineers
British Astronomical Association (and its *Journal*)
British Institute of Radiology (and its *Journal*)
British Mycological Society (and its *Transactions*)
Mathematical Gazette
Journal of Pathology and Bacteriology
Science Abstracts
Belfast Association of Engineers
Imperial Institute
West of Scotland Iron and Steel Institute
Institution of Public Health Engineers
British Bryological Society
Lister Institute of Preventive Medicine
London School of Tropical Medicine
Institute of Refrigeration
Battersea Polytechnic Institute
University College Reading
Malacological Society
Society of Physiotherapy
Northampton Institute, London
Natural Resources Institute
Scottish Marine Biological Association

Institution of Building Services Engineers
Royal Technical College, Salford
University College Sheffield
Photographic Review
Medical Review

Appendix C

Journals Related to Physical Chemistry

Abbreviations:

adv = advances
ann = annals, annales, annalen, etc.
AS = academy of sciences
ass = association, etc.
bull = bulletin
ch = chemistry, chemical, chimica, etc .
inst = institute, etc.
intl = international
izv = izvestiya
j = journal
p = proceedings
prog = progress, etc.
res = research
rev = review, revue, revista, etc.
soc = society, societe, etc.
univ = university, etc.
z = zeitschrift
zh = zhurnal

All journals are general chemistry except:

* physical chemistry
** subdivisions

1828	j prakt ch - german
1832	liebig's ann ch - german
1836	p roy irish acad, ch
1858	bull soc ch france
1868	ch berichte - german
1871	gazz ch italiana
1876	ch listy - czech
1876	ch zeitung - german
1879	j amer ch soc
1880	monatshefte ch - german
1882	rec trav ch - holland
1887	* z phys ch - german
1887	bull soc ch belge
1889	kemisk tidskr - sweden
1892	z anorg allgemein ch - german
1895	hungarian j ch
1896	* j phys ch - usa
1897	* ber phys ch - german
1902	** j electroch soc - usa
1903	* j ch et phys ch - france
1903	ann quimica - spain
1903	ch magazine - holland
1903	ch weekblad - holland
1906	** colloid & polymer sci - german
1909	** prog colloid & polymer sci - german
1912	ann, ass quim argentina
1913	kjemi - norway
1914	ann di ch - italy
1914	ch bull - usa
1916	** catalyst - usa
1917	przemysl ch - poland
1918	helvetica ch acta - switz
1919	collec czech ch commun
1921	afinidad - spain
1924	ch rev - usa
1924	j indian ch soc
1926	bull ch soc japan
1927	croat ch acta
1929	j inst ch - india
1929	canadian j ch

1930 * zh fizicheskoi khim - ussr
1930 j serbian ch soc
1931 zh obshchei khim - ussr
1932 uspekhi khimii - ussr
1933 * j ch phys - usa
1933 acta ch sinica - china
1933 ch bull - china
1934 p indian AS, ch
1934 bull soc ch peru
1935 * kolloidnyi zh - ussr
1936 AS SSSR, ch
1942 ann AS finland, ch
1946 * j colloid sci - usa
1946 ann univ marie-curie, ch - poland
1946 j korean ch soc
1946 prog ch - korea
1947 AS kazakh, ch
1947 ch zvesti - slovakia
1947 ch rundschau - switz
1947 wiadomosci ch - poland
1948 kyushu univ, ch - japan
1948 chemik - poland
1948 s african j ch
1948 australian j ch
1949 rev ch - rumania
1949 bull chilean ch soc
1950 adv in ch - usa
1950 hemijski pregled - serbia
1951 israel j ch
1951 acta ch hungarica
1951 * ann rev phys ch - usa
1954 bull slovenian ch soc
1954 j chinese ch soc
1954 polytech lodzka, ch - poland
1955 folia ch - poland
1956 AS estonia, ch
1956 rev roumaine ch
1957 * j molecular spectroscopy - usa
1957 rev soc quimica mexico
1957 uzbek ch j

1958 * adv ch phys - usa
1958 * molecular phys - uk
1958 egyptian j ch
1959 azerbaijan ch j
1959 * electroch acta - holland
1960 israel ch
1960 pure & applied ch - intl
1960 chemia - poland
1961 kimika - philippines
1961 * prog reaction kinetics - uk
1961 AS latvia, ch
1962 * theor ch acta - german
1962 * j catalysis - usa
1963 * adv phys organic ch - usa
1963 AS siberia, ch
1963 * prog phys organic ch - usa
1963 * adv photoch - holland
1964 * adv quantum ch - usa
1964 nigata univ, ch - japan
1965 * kataliz - ukraine
1965 ch commun - uk
1965 topics in current ch - intl
1965 AS beloruss, ch
1965 teoret & eksper khim - ukraine
1966 mendeleev ch j - ukraine
1966 armenian ch j
1966 * internatl j quantum ch - usa
1967 * ch phys letters - holland
1967 * j molecular struc - holland
1967 * adv colloid sci - holland
1967 * catalysis rev - usa
1967 kyoto univ, ch res bull - japan
1968 izv ch - bulgaria
1968 accts ch res - usa
1968 * j ch kinetics - usa
1969 * comprehen ch kinetics - holland
1969 * j ch thermodynamics - usa
1969 rev quimica - cuba
1969 * j molecular liquids - holland
1970 * photochem - uk

1970 * thermoch acta - holland
1970 rev quimica - mexico
1971 ch scripta - sweden
1972 * j solution ch - usa
1972 ch letters - japan
1972 ch soc rev - uk
1972 j ch soc japan
1973 * ch phys - holland
1973 actualite ch - fr
1974 * reaction kinetics - hungary
1975 * j molecular catalysis - holland
1976 * j nonequilib thermodyn - holland
1976 eclectica ch - brazil
1976 turkish j ch
1976 j iraqi ch soc
1977 j ch res - uk
1977 nouveau j ch - france
1978 * fluid phase equilibria - holland
1978 * spectrochem acta rev - uk
1979 ch international
1979 j ch soc pakistan
1980 * colloids & surfaces - holland
1980 bull korean ch soc
1980 * j dispersion sci - usa
1981 * theochem - holland
1981 * internatl rev phys ch - intl
1981 * catalysis sci & tech - intl
1982 * aerosol sci & tech - usa
1982 * spectroscopy - intl
1983 chinese j ch
1985 * langmuir - usa
1987 * catalysis today - holland
1988 * catalysis letters - switz
1989 * spectroscopy internatl - usa
1990 * adv catalysis - usa
1990 * structural ch - german
1990 * vibrational spectroscopy – holland

Appendix D

Databases Relevant to Environmental Issues, 1990

The database title, its country of origin, and its type (bib = bibliography, num = numerical):

Acid Rain - USA - bib
Acidoc - Canada - bib
AFEE - France - bib
AGRIS - international - bib
Agrochemicals Handbook - UK - text/num
AMILIT - Sweden - bib
APTIC - USA - bib
AQUALINE - USA - bib
Aquatic Information Retrieval - USA - text/num + bib
Aquatic Science and Fisheries Abstract - international - bib
BNA Occupational Safety - USA - full text
BNA Toxics - USA - full text
ChemicalActivity - USA - referral
Chemical Exposure - USA - text/num + bib
Chemical Hazards - USA - text/num
Chemical Regulation Reporter - USA - full text
Chemical Regulations and Guidelines - USA - bib
Chemical Safety Data - USA - full text
Chemical Safety Newsbase - UK - bib + referral
Chemical Substances Control - USA - full text
Cheminfo - Canada - text/num
Chemlist - USA - referral

CISDOC - international - bib
CTCP - USA - text/num
DETEQ - Germany - referral
DoE Energy - USA - bib
ECDIN - international - text/num + bib
EMTOX - Netherlands - bib
ENREP - international - referral
Enviroline - USA - bib
Environment Reporter - USA - full text
Environment Week - USA - full text
Environmental Bibliography - USA - bib
Environmental Fate - USA - text/num
Environmental Fate Data - USA - text/num + bib
Environmental Health News - USA - full text
Environmental Law Reporter - USA - full text
Environmental Mutagens - USA - bib
GEOLINE - Germany - bib
GEOS - Germany - referral
Hazardline - USA - text/num
Hazardous Materials - USA - full text + referral
Hazardous Waste News - USA - full text
HAZINF - Canada - text/num
Health and Safety Abstract - UK - bib
HSDB - USA - text/num
HSELINE - UK - bib
Industrial Health and Hazards - USA - full text + bib
Industry Response - USA - full text
INFOTOX - Canada - text/num
INIS - international - bib
IRIS - USA - text/num + referral
ISST - Canada - bib + referral
Job Safety and Health - USA - full text
LEXIS Environmental Library - USA - full text
MALLIN - USA - referral
Medical Waste News - USA - full text
Multinational Environmental Outlook - USA - full text
National Waste Exchange - USA - referral
NEDRES - USA - referral + bib
NIOSHTIC - USA - bib
Normes et repertoires - Canada - bib

NTIS - USA - bib
Nuclear Fuel - USA - full text
Nuclear Science Abstract - USA - bib
Nuclear Waste News - USA - full text
Occupational Safety and Health - USA - full text
OHM-TADS - USA - text/num
OHM-MSDS - USA - text/num
PESTDOC - UK - bib
Pollution Abstract - UK - bib
REM - international - numeric
Report on Defense Plant Wastes - USA - full text
RTECS - USA - text/num
SCARABEE - France - bib
SERIX - Sweden - bib + referral
SIGLE - international - bib
Solid Waste Report - USA - full text
Suspect Chemicals - USA - text/num + referral
Toxic Chemicals Release - USA - text/num
Toxic Materials News - USA - full text
Toxic Materials Transport - USA - full text
Toxics News - USA - bib + full text
TOXLINE - USA - bib
TOXLIT - USA - bib
Transportation of Dangerous Goods - Canada - text/num
TSCA - USA - referral + text/num
TSCATS - USA - bib + text/num
Umweltforschungs - Germany - bib
Umweltliteratur - Germany - bib
Waste Management - international - bib
Waste-to-Energy - USA - full text
Water Resources Abstract - USA - bib
Waternet - USA - bib

Appendix E

French Centre National de la Recherche Scientifique

Some Research Institutes, 1990

MATHEMATICS AND PHYSICS

Theoretical physics
Molecular photophysics
Strong magnetic fields
Quantum optics
Physics of solids
Magnetism
Physics of materials
Very low temperatures
Mechanical and thermodynamic properties
Imperfect crystals
Crystallography
Electron optics
Crystal growth
Neutron spectrometry
Use of electromagnetic radiation
Heavy ions
Electron microscopy
Microelectronics
Plasmas
Polarized light
Dissymetric organic structures
Nonlinear interactions
High-temperature physics

Aerothermics
Materials engineering
Physics of oscillators
Dielectric materials
Engineering at high pressure
Physics of discharges
Atomic clock
Physics of interfaces
Semiconductors
Lasers
Turbulence
Nuclear spectrometry
Heavy-ion accelerator
Linear accelerator

CHEMISTRY

Catalytic reactors
Combustion
Metallurgical chemistry
Chemistry of solids
Natural substances
Gas-solid interactions
Irradiation
Interface electrochemistry
Plant macromolecules
Microcalorimetry
Infrared spectrochemistry
Coordination chemistry
Phosphorus chemistry
Organometallics
Thermostructural composites
Molecular modeling
Thermostable polymers
Stereochemistry

LIFE SCIENCES

Physiological optics
Molecular biophysics

Biopolymers
Bio-active products
Macromolecular biochemistry
Laboratory animals
Wild animals
Cellular genetics
Cancer
Bacterial chemistry
Embryology
Enzymology
Virus genetics
Neurochemistry
Nutrition
Physiological regulation
Nerve physiology
Evolutionary genetics
Immunology
Neurosensorial physiology
Cellular biology
Environmental physiology
Viral oncology
Photosynthesis
Phytosociology
Immunopharmacology
Genetic polymorphism
Genetic recombination
Retroviruses
Interferons
Photosynthetic membranes
Phytotron

EARTH AND SPACE SCIENCES

Aeronomy
Spatial astronomy
Astrophysics
Solar energy
Geophysics
Glaciology
Physics of the environment

Dynamic meteorology
Petrography
Feeble radioactivity
Surface geochemistry
Minerals
Geomorphology
Tropical geography
Eco-geography

Appendix F

Science, Research and History: B.C. Vickery, 1959

Homer, Sophocles, Plato, Omar, Chaucer, Cervantes, Shakespeare, Bunyan, Swift, Burns, Shelley, Carlyle, Dickens, Hugo, Tolstoi, Conrad . . . the influence of such writers grows from year to year. As literacy spreads in each generation the classics of literature win a greater number of readers.

How different is the fate of the classical texts of science! Archimedes, Galen, Ptolemy, Avicenna, Grosseteste, Vesalius, Galileo, Newton, Lavoisier, Liebig, Pasteur, Mendeleev . . . who reads them now, except the professional historian, the occasional student, the antiquarian? In science, it is widely held that current knowledge supersedes all knowledge of the past. All useful earlier science is absorbed in that of the present; what has been left out are only the mistaken ideas of ignorant predecessors. There is no point in studying the classics of a bygone age. In the words of Henry Ford, "History is bunk."

This opinion, as I have said, is widely held, although a knowledge of the history of science, and a conviction of its value, are to be observed among many eminent scientists. It is true that the empirical facts established and confirmed by earlier workers are usually to be found in current reference works, easier to consult than a century-old original. It is also true that the classic texts enshrine many ideas that have since been shown to be false—the proportion of error and fancy to truth increasing as we go backwards in time. It is further true that science is in its essence revolutionary and progressive, advancing by breaking with old tradition. Yet, in spite of these truths, the everyday practice of science itself demonstrates the use of history in research.

What, after all, is a literature survey but a piece of historical research, undertaken to establish what is known about a certain phenomenon, and what methods have been used to investigate it? A good deal of the working scientist's time is spent on such research, and it is no easy task: as Lord Rayleigh remarked over a hundred years ago, "rediscovery in the library may be a more difficult and uncertain process than the first discovery in the laboratory." Even the technique of the search may recall that of the literary commentary: we may look up an encyclopedia or review to see whom the author thought worth quoting; then examine the cited papers to see whom they quote; and so on back into the past.

The usual scientific literature search, of course, stretches back only into the recent past. Many analyses of citations in scientific papers have been made, and a fairly constant pattern emerges: more than half the citations are less than ten years old, another quarter are between ten and twenty years old, less than ten percent are over thirty years old. In a complete bibliography of a current research topic, the proportion of older references is greater, but not considerably so. In the taxonomic sciences, papers over fifty years old may constitute twenty to thirty percent of the total. In his historical search for facts and methods, the scientist is usually justified in consulting only the recent past.

I have heard librarians conclude from this that no periodical article more than, say, thirty years old is of any importance for research. To hold this opinion is to regard science as the simple amassing of facts and methods, which all eventually get recorded in reference books. Recourse to the original sources for the last thirty years is then only necessary because of the time lag in compiling reference books. This view neglects the fundamental importance of imagination and ideas in the development of science. Science is essentially a process, stretching through time, and ideas formulated long ago at an early stage of the process may fruitfully interact with facts discovered only yesterday. The atomic theory of chemistry is a classic example.

The value to the research worker of breadth of interest is often acknowledged. But, as George Sarton said in his *Guide to the History of Science*, "when we contemplate the universe we may adopt one of two points of view . . . the side-by-sidedness of things or their one-after-anotherness. It would be misleading to say that the second point of view is exclusive to the historian, and the first to the naturalist. . . . In reality, both points of view are necessary and complementary." If a broad understanding of the panorama of contemporary knowledge is valuable, so also is an understanding of the development of knowledge down the centuries.

The reason has already been suggested: confirmed facts and tested methods get into the reference books, but ideas may often be lost, if they do not strike a responsive chord in the scientific community at the time of their first expression. We have often—perhaps too often—been regaled with the story of Mendel's work on the heredity of peas, "lost" for over thirty years because, it is alleged, it was published in an obscure journal. However, the work did not remain unknown at its first appearance: Mendel reported it to Nägeli, one of the foremost botanists of the nineteenth century, but failed to convince him of its significance. Mendel's ideas were lost, not simply by obscurity of publication, but because they did not fit in with the scientific notions of his contemporaries. One valid reason for the search through historical literature is to rediscover such forgotten ideas, discarded as erroneous at one time, but regaining vitality in the light of new facts.

On the other hand—and this is perhaps truer of disciplines less firmly based on fact than the natural sciences—there is another danger against which historical knowledge can safeguard us. In the more speculative fields of study, ideas that have already proved unfruitful are continually being suggested anew, leading as continually to failure and disappointment. An example occurs in the history of the classification of science during the nineteenth century. Throughout this period, a succession of scholars tried to divide up the sciences by a series of subjective dichotomies—speculative sciences and descriptive sciences, pure sciences and mixed sciences, formal and empirical, normative and physical, abstract and concrete, fundamental and dependent, general and special, exact and synoptic. Such terms were used in repeated—and contradictory—attempts to group the sciences. The fact that the attempt was so often repeated, and that earlier terminology was continually discarded as unsuccessful, suggests to the historian that the method itself—the subjective method of division—was false. Better understanding of the history of their problem might have made this conclusion evident sooner to the classifiers themselves.

Historical knowledge, therefore, helps the researcher to retrieve forgotten but fruitful ideas, and to avoid resurrecting unfruitful ones. It has also a broader function. A knowledge of the history of a subject generates understanding of the manner in which it develops. We find, for example, that in science the most active points of advance have often been those where previously separate fields of study have come together—e.g., the science of communications today. As a result of this, many major advances in a particular science have come from individuals who have had relatively little experience in that science. A sense of history, an understanding of the strategy and tactics of scientific advance, helps the researcher in both the choice and the organization of his work.

For the librarian who serves the researcher, there are obvious morals to be drawn. If we neglect to stock—and to disseminate—the older literature of our subject, and histories of it, then we may unwittingly deprive our users of a stimulus and a source of valuable ideas. Again, in ourselves studying—however superficially—the subjects covered in our libraries, we may give more valuable reference service if we acquire historical as well as contemporary knowledge.

There is a further moral to be drawn. Librarianship and documentation have long had some pretensions to science themselves, and the volume of research in these areas is beginning to increase. A research tradition is in the course of being built up. To guide this work in the most useful directions, an historical understanding of our own professional problems is needed. When money is being offered for new gadgets, new systems, new surveys, the claims of historical research in librarianship and documentation should not be overlooked.

One last point. "All that mankind has done, thought, gained or been," wrote Thomas Carlyle, "is lying as in magic preservation in the pages of books." We whose business is books well know the abiding pleasure and inspiration to be derived from the study of "all that mankind has done." History not only serves research, it enlivens the heart and mind of man.

Appendix G

Development of the Internet

The Internet is a worldwide network of networks, providing standard ways of linking together, over the telephone system, an immense variety of computer systems, ranging from the largest supercomputers to desktop personal computers.

The Internet may be said to have "emerged" with the definition of an Internet protocol (IP) for data transmission in 1982, going more decisively public with the construction of a new "backbone" (NSFnet) in 1987, but it had a long prehistory. The foundation was laid by the development of packet-switched communications and time-shared computers in the early 1960s. By 1965, project INTREX at MIT had come up with the concept of "the online intellectual community." In the later sixties there were several parallel developments: (1) the construction of online database services using the public telephone system (at first with restricted access, but soon becoming publicly available); (2) the formation of private "bulletin boards" accessible over the telephone system; and (3) the creation of specialized computer communication systems such as Arpanet, used for government and academic access to computer services and electronic mail—computers in these nets also provided application software and research data banks.

During the early seventies, public network services such as Tymnet and Telenet were established, and direct overseas telephone dialing was introduced, giving wider possibilities for both online access and electronic mail. The development of personal computers soon provided "intelligent" terminals for data communications. Work began on the standardization of communications protocols—e.g., X.25 and UUCP. In the later seventies, new specialist networks were created, for example Usenet, CompuServe, Euronet/Diane.

From 1982, Arpanet began to evolve into the overarching, publicly available service, the Internet, later to be based on the new NSFnet backbone. In the 1990s, Internet facilities were developed: Archie, WAIS, Gopher, Veronica, the World Wide Web, new search engines. A governing Internet Society was formed in 1992.

In principle, any facility on any computer on any network is available through the Internet. In practice, of course, private or restricted networks may not (yet) have established links to the Internet, either because they do not see that it offers advantages over their present arrangements, or because of foreseen disadvantages (e.g., possible lack of security). Some details of this thirty-year development are set out below.

1960s Packet switching
1961 Time-sharing project MAC at Massachusetts Institute of Technology (MIT)
1962 Online Technical Information Project at MIT
1964 Space Documentation Service set up in Europe (later to become ESA/IRS)
1965 Project Intrex at MIT: "the online intellectual community"
1966 Arpanet commissioned by U.S. Department of Defense
1969 Abridged Index Medicus by Teletypewriter Exchange Network
1970 Electronic mail starts
1970 Arpanet nodes use Network Control Protocol
1971 Tymnet starts
1971 MEDLARS online
1971 Direct overseas dialing introduced
1972 DIALOG as a commercial service
1972 InterNetworking Group set up to establish protocols
1973 Arpanet connections to Europe
1974 DowJones news/retrieval service
1975 Telenet opened by Bolt Beranek Newman (later Sprintnet)
1975 First personal computers
1975 Early bulletin boards
1976 Unix-to-Unix Copy Protocol developed by AT&T Bell Labs
1976 X.25 established by CCITT as connection standard
1977 Theorynet (computer-science network) developed at University of Wisconsin
1977 STN International

1979	Usenet established using UUCP
1979	CompuServe starts
1979	Euronet/Diane database network set up
1980	DATASTAR established
1981	PSS in U.K.
1981	Bitnet started at City University of New York
1981	First online front ends and search aids
1982	INWG creates TCP/IP for Arpanet
1982	Eunet established
1983	MCI offers email to public
1984	Domain Name Server introduced on Arpanet
1984	Compulink Information Exchange
1985	Easynet
1985	Byte Information Exchange
1986	NSFnet created
1987	NSFnet contracts with IBM, MCI, and Merit Network to manage NSFnet backbone
1989	America Online
1990	Archie released
1991	Commercial Internet Exchange
1991	WAIS released by Thinking Machines Corporation
1991	Gopher released by University of Minnesota
1992	Veronica developed at University of Nevada
1992	Internet Society formed
1992	World Wide Web designed at CERN
1992	Commercial Internet service providers appear
1993	InterNIC created by NSF
1993	MOSAIC WWW browser released by NCSA
1994	Net search engines are developed
1995	Netscape WWW browser released
1995	Yahoo net directory
1996	Intelligent agent for WWW searching
1997	Electronic journals on the Web

Bibliography

Ackoff, R.L. and M.H. Halbert. *An Operations Research Study of the Scientific Activity of Chemists.* Cleveland: Case Institute of Technology, 1958.

Allen, D.E. *The Naturalist in Britain: A Social History.* London: Allen Lane, 1976.

Allen, T. *Managing the Flow of Scientific and Technical Information.* Cambridge, Mass.: MIT School of Management, 1966.

Andrews, E. *History of Scientific English.* New York: Smith, 1947.

Arber, A. *Herbals, Their Origin and Evolution.* London: Cambridge University Press, 1938.

Armstrong, C.J. and J.A. Large, ed. *Manual of Online Search Strategies.* Aldershot, Hants.: Gower, 1988.

Armytage, W.H. *A Social History of Engineering.* London: Faber, 1970.

———. *The Rise of the Technocrats.* London: Routledge, 1965.

Ashford, J. and P. Willett. *Text Retrieval and Document Databases.* London: Chartwell-Bratt, 1988.

Auger, C.P., ed. *Use of Reports Literature.* London: Butterworths, 1975.

Babbage, C. *On the Economy of Machinery and Manufactures.* London, 1832.

Bairoch, P. "International Industrialisation Levels from 1750 to 1980." *Journal of European Economic History* 11 (1982): 281–96.

Barnes, S.B. "The Editing of Early Learned Journals." *Osiris* 1 (1936): 155–72.

Barr, D. *Book and Serial Publishing Trends, 1951–80.* London: National Libraries ADP Study, 1971.

Barry, G., ed. *Communication and Language.* London: MacDonald, 1965.

Bates, R.S. *Scientific Societies in the United States.* Oxford: Oxford University Press, 1965.

Bell, D. *The Coming of Post-Industrial Society.* Harmondsworth: Penguin, 1973.

Ben-David, J. *The Scientist's Role in Society.* Englewood Cliffs, N.J.: Prentice-Hall, 1971.

Beresford, M. *New Towns of the Middle Ages*. London: Lutterworth, 1967.

Bernal, J.D. *Science in History*. London: Watts, 1965.

———. *Science and Industry in the Nineteenth Century*. London: Routledge, 1953.

———. *Social Function of Science*. London: Routledge, 1939.

Besterman, T. *The Beginnings of Systematic Bibliography*. Oxford: Oxford University Press, 1935.

———. *World Bibliography of Bibliographies*. Lausanne: Societas Bibliographica, 1965–66.

Blunt, W. *The Art of Botanical Illustration*. London: Collins, 1950.

Bolgar, R.R. *The Classical Heritage and Its Beneficiaries*. London: Cambridge University Press, 1954.

Bolton, H.C. *Catalogue of Scientific and Technical Periodicals from 1665 to 1895*. Washington: Smithsonian Institution, 1897.

Boorstin, D.J. *The Discoverers*. Harmondsworth: Penguin, 1983.

Borgman, C.L., ed. *Scholarly Communication and Bibliometrics*. New York: Sage, 1992.

Bowen, J. *History of Western Education*. London: Methuen, 1972–5.

Bowler, P.J. *The Environmental Sciences*. London: Fontana, 1992.

Boyer, C.J. *The Doctoral Dissertation as an Information Source*. Metuchen, N.J.: Scarecrow Press, 1973.

Bradford, S.C. "Sources of Information on Specific Subjects." *Engineering* 137 (1934): 85–86.

Braudel, F. *History of Civilizations*. London: Penguin Press, 1994.

Brodman, E. *The Development of Medical Bibliography*. Chicago: Medical Library Association, 1954.

Brown, H. *Scientific Organizations in Seventeenth Century France*. Baltimore: Williams and Wilkins, 1934.

Brown, J. "Guild Organisation and the Instrument-Making Trade, 1550–1830." *Annals of Science* 36 (1979): 1–34.

Brown, R.W. *Composition of Scientific Words*. Baltimore: The Author, 1954.

Buckle, H.T. *History of Civilisation in England*. London: Routledge, 1904.

Bud, R. and D.J. Warner, ed. *Instruments of Science*. London: Garland, 1998.

Burton, R.E. and R.W. Kebler. "The Half-Life of Some Scientific and Technical Literatures." *American Documentation* 11 (1960): 18–30.

Bush, V. "As We May Think." *Atlantic Monthly* 176 (1945): 101–8.

———. *Science the Endless Frontier*. Washington: U.S. Government Printing Office, 1945.

Bynum, W.F. et al. *Dictionary of the History of Science*. London: MacMillan, 1981.

Cardwell, D.L. *The Organisation of Science in England*. London: Heinemann, 1972.

Carter, T.F. *The Invention of Printing in China and Its Spread Westward*. New York: Ronald Press, 1955.

Chambers Biographical Dictionary. Edinburgh: Chambers, 1974.

Chappell, W. *A Short History of the Printed Word*. London: Deutsch, 1972.

Cherry, C. *World Communication: Threat or Promise*. London: Wiley-Interscience, 1971.

Childe, V. Gordon. *What Happened in History*. Harmondsworth: Penguin, 1964.

Chisholm, M. *Modern World Development: A Geographical Perspective*. London: Hutchinson, 1982.

Ciba Foundation. *Communication in Science*. London: Churchill, 1967.

Cipolla, C.M. *European Culture and Overseas Expansion*. Harmondsworth: Pelican, 1970.

———. *Literacy and Development in the West*. Harmondsworth: Penguin, 1969.

Cole, F.J. and N.B. Eales. "The History of Comparative Anatomy." *Science Progress* 11 (1917): 578–96.

Collison, R. *Encyclopedias: Their History throughout the Ages*. New York: Hafner, 1966.

Committee on Scientific and Technical Communication. *Scientific and Technical Communication*. Washington: U.S. National Academy of Sciences, 1969.

Condit, L. "Bibliography in Its Prenatal Existence." *Library Quarterly* 7 (1937): 564–76.

Corsi, P. and P. Weindling. *Information Sources in the History of Science and Medicine*. London: Butterworths, 1983.

Cowan, R.S. *Social History of American Technology*. Oxford: Oxford University Press, 1997.

Crombie, A.C. *Augustine to Galileo*. Harmondsworth: Penguin, 1969.

Crosland, M.P. *Historical Studies in the Language of Chemistry*. London: Heinemann, 1962.

Crowther, J.G. *Social Relations of Science*. London: Dufour, 1941.

Crystal, D., ed. *Cambridge Encyclopedia of Language*. London: Cambridge University Press, 1987.

Cuadra, C., ed. *Directory of Online Databases*. New York: Cuadra/Elsevier, 1990.

———. *Directory of Portable Databases*. New York: Cuadra/Elsevier, 1990.

Curwen, H. *A History of Booksellers*. London: Chatto and Windus, 1873.

Dahl, S. *History of the Book*. Metuchen, N.J.: Scarecrow Press, 1958.

Daiches, D. and A. Thorlby, ed. *Literature and Western Civilisation*. London: Aldus, 1972–6.

Daumas, M., ed. *Histoire de la Science*. Paris: Gallimard, 1957.

———. *History of Technology and Invention*. London: Murray, 1979–80.

Davinson, D.E. *Theses and Dissertations as Information Sources*. London: Bingley, 1977.

Dembowska, M. *Documentation and Scientific Information*. Warsaw: National Science Foundation, 1968.

Dempsey, L., ed. *Bibliographic Access in Europe*. Aldershot, Hants.: Gower, 1990.

———. *Libraries, Networks and OSI*. Westport, Conn.: Meckler, 1991.

Derry, T.K. and T.I. Williams. *Short History of Technology*. London: Oxford University Press, 1960.

DeVore, P.W. *Technology: An Introduction*. Worcester, Mass.: Davis, 1980.

Dicken, P. *Global Shift*. London: Paul Chapman, 1992.
Dicken, P. and P.E. Lloyd. *Modern Western Society*. London: Harper and Row, 1981.
Dietz, F.C. *Economic History of England*. New York: Holt, 1942.
Diringer, David. *The Hand-Produced Book*. London: Hutchinson, 1953.
Dolby, R.G.A. "The Transmission of Science." *History of Science* 15 (1977): 1–43.
Dossett, P., ed. *Handbook of Special Librarianship and Information Work*. London, Aslib, 1992.
Durbin, P.T., ed. *A Guide to the Culture of Science, Technology and Medicine*. London: MacMillan, 1980.
Earle, P. and B.C. Vickery. "Subject Relationships in Science/Technology Literature." *ASLIB Proceedings* 21 (1969): 237–43.
East, W.G. *The Geography behind History*. London: Nelson, 1965.
Edge, D. "Quantitative Measures of Communication in Science: A Critical Review." *History of Science* 17 (1979): 102–34.
Egret, D. and M.A. Albrecht, ed. *Information and Online Data in Astronomy*. London: Kluwer, 1995.
Eisenstein, E.L. *The Printing Press as an Agent of Change*. London: Cambridge University Press, 1979.
Encyclopaedia Britannica, various editions.
Farrington, B. *Francis Bacon: Philosopher of Industrial Science*. London: Lawrence and Wishart, 1951.
Febvre, L. and H.J. Martin. *The Coming of the Book*. London: Verso, 1976.
Ferchl, F. and A. Sussenguth. *Pictorial History of Chemistry*. London: Heinemann, 1939.
Ferguson, J. *The Heritage of Hellenism*. London: Thames and Hudson, 1973.
Feuer, L.S. *The Scientific Intellectuals*. New York: Basic Books, 1965.
Fisher, J.S., ed. *Geography and Development*. New York: MacMillan, 1992.
Flocon, A. *L'univers des livres*. Paris: Hermann, 1961.
Forbes, R.J. *Man the Maker*. London: Constable, 1950.
Forester, T., ed. *The Microelectronics Revolution*. Oxford: Blackwell, 1980.
Freeman, R.R. "Current Research Information in the USA." *Journal of Information Science* 20 (1994): 556–62.
Garfield, E. *Citation Indexing*. New York: Wiley, 1979.
Garfield, E. and A. Welljams-Dorof. "Language Use in International Research." *Annals of American Academy of Political and Social Science* 511 (1990): 10–22.
Garvey, W.D. *Communication, the Essence of Science*. Oxford: Pergamon, 1979.
Gibbon, E. *Decline and Fall of the Roman Empire*. London, 1776–88.
Giddens, A. *Sociology*. Oxford: Polity Press, 1993.
Gille, B. *Histoire des techniques*. Paris: Gallimard, 1978.
Gillispie, C.C., ed. *Dictionary of Scientific Biography*. New York: Scribner, 1970–1980.

Goodman, D. and C. Russell, ed. *The Rise of Scientific Europe 1500–1800*. London: Hodder and Stoughton, 1991.

Gralewska-Vickery, A. "Communication and Information Needs of Earth Science Engineers." *Information Processing and Management* 12 (1976): 251–82.

Gralewska-Vickery, A. and H. Roscoe. *Earth Science Engineers: Communication and Information Needs*. London: Imperial College, 1975.

Griffith, B.C. et al. "Aging of Scientific Literature: A Citation Analysis." *Journal of Documentation* 35 (1979): 179–96.

Grogan, D. *Science and Technology: An Introduction to the Literature*. London: Bingley, 1982.

Guerlac, H. *Science in Western Civilisation: A Syllabus*. New York: Ronald Press, 1952.

Guppy, H. "Human Records: A Survey of Their History from the Beginnings." *John Rylands Library Bulletin* 27 (1942): 182–222.

Hagstrom, W.O. *The Scientific Community*. New York: Basic Books, 1965.

Hall, J.L. *On-line Information Retrieval Sourcebook*. London: ASLIB, 1977.

Hall, P. and P. Preston. *The Carrier Wave: New Information Technology and the Geography of Innovation 1846–2003*. London: Unwin Hyman, 1988.

Hamilton, H. *History of the Homeland*. London: Allen and Unwin, 1947.

Hannaway, O. *The Chemists and the Word*. Baltimore: Johns Hopkins University Press, 1975.

Harrison, J.F.C. *The Common People*. London: Fontana, 1984.

Haskins, C.H. *Studies in the History of Medieval Science*. Cambridge, Mass.: Harvard University Press, 1924.

———. *The Renaissance of the Twelfth Century*. Cambridge, Mass.: Harvard University Press, 1927.

Hastings, R. *The Universities of Europe in the Middle Ages*. London: Oxford University Press, 1936.

Havelock, R.G. et al. *Comparative Study of the Literature on the Dissemination and Utilisation of Scientific Knowledge*. Ann Arbor: Michigan University Press, 1969.

Hellemans, A. and B. Bunch. *The Timetables of Science*. New York: Simon and Schuster, 1988.

Hepworth, M. *Geography of the Information Economy*. London: Belhaven, 1989.

Herner, S. "Brief History of Information Science." *Journal of American Society for Information Science* 35 (1984): 157–63.

Hessel, A. *History of Libraries*. Metuchen, N.J.: Scarecrow Press, 1955.

Hilton, A.M. *Logic, Computing Machines and Automation*. Washington: Spartan Books, 1963.

Hjerppe, R. *An Outline of Bibliometrics and Citation Analysis*. Stockholm: Royal Institute of Technology Library, 1978.

Hogben, L. *From Cave Painting to Comic Strip*. London: Parrish, 1960.

———. *The Vocabulary of Science*. London: Heinemann, 1969.

Holmstrom, J.E. *Records and Research in Engineering and Industrial Science*. London: Chapman and Hall, 1956.

Holton, G. "Scientific Research and Scholarship." *Daedalus* 91 (1962): 362–99.

Hough, J.N. *Scientific Terminology*. New York: Rinehart, 1953.

Hufbauer, K. *The Formation of the German Chemical Community*. Berkeley: University of California Press, 1982.

Huff, T.E. *Rise of Early Modern Science*. London: Cambridge University Press, 1993.

Hulme, E.W. *Statistical Bibliography in Relation to the Growth of Modern Civilization*. London: Grafton, 1923.

Illinois Institute of Technology. *Technology in Retrospect and Critical Events in Science*. Washington: National Science Foundation, 1968.

Information. *Scientific American* 215, nr. 3 (1966): 1–316.

Information Science in America. *Bulletin of American Society for Information Science* 2, nr. 8 (1976): 1–60.

Inkster, I. *Science and Technology in History*. London: MacMillan, 1991.

Irwin, R. *Heritage of the English Library*. New York: Hafner, 1964.

Isis. Critical Bibliographies. Philadelphia: History of Science Society, 1912 on.

Iwinski, M.B. "La statistique internationale des imprimees." *Bulletin de l'internationale institut de bibliographie* 16 (1911): 1–139.

Jackson, S.L. *Libraries and Librarianship in the West*. New York: McGraw-Hill, 1974.

Jaeger, E.C. *Sourcebook of Biological Terms*. Springfield, Ill.: Thomas, 1955.

Jaffe, B. *Men of Science in America*. New York: Simon and Schuster, 1944.

Johnson, E.D. *Communication: A Concise Introduction to the History of the Alphabet, Writing, Printing, Books and Libraries*. Metuchen, N.J.: Scarecrow Press, 1955.

———. *History of Libraries in the Western World*. Metuchen, N.J.: Scarecrow Press, 1965.

Kaplan, N. "Sociology of Science," in *Handbook of Modern Sociology*. Chicago: Rand McNally, 1964.

Kent, A., ed. *Encyclopedia of Library and Information Science*. New York: Dekker, 1968 on.

Kenyon, F.D. *Books and Readers in Ancient Greece*. Oxford: Clarendon Press, 1932.

Klebs, A.C. "Incunabula scientifica et medica." *Osiris* 4 (1938): 1–359.

Knox, P. and J. Agnew. *The Geography of the World Economy*. London: Arnold, 1989.

Kochen, M., ed. *The Growth of Knowledge*. New York: Wiley, 1967.

Kranzberg, M. and C.W. Pursell. *Technology in Western Civilization*. Madison: University of Wisconsin Press, 1967–8.

Krige, J., ed. *Science in the Twentieth Century*. Amsterdam: Harwood, 1997.

Krommer-Benz, M. *World Guide to Terminological Activities*. Munich: Verlag Dokumentation, 1977.

Kronick, D.A. *A History of Scientific and Technical Periodicals*. Metuchen, N.J.: Scarecrow Press, 1976.

Kuhn, T.S. *The Structure of Scientific Revolutions*. Chicago: University of Chicago Press, 1970.

Lambert, S. and S. Ropiequet, ed. *CD/ROM: The New Papyrus*. Redmond: Microsoft Press, 1986.

Lancaster, F.W. *Vocabulary Control for Information Retrieval*. Washington: Information Resources Press, 1986.

Lancaster, F.W. and E.G. Fayen. *Information Retrieval On-line*. Los Angeles: Melville, 1973.

Landes, D. *The Unbound Prometheus*. London: Cambridge University Press, 1969.

Langdale, J.V. "The Geography of International Business Telecommunications." *Annals of Association of American Geographers* 79 (1989): 501–22.

Lenski, G. and J. Lenski. *Human Societies*. New York: McGraw-Hill, 1978.

Liebesny, F. *Mainly on Patents*. London: Butterworths, 1972.

Lightman, B., ed. *Victorian Science in Context*. Chicago: University of Chicago Press, 1997.

Lilley, D.B. and R.W. Trice. *A History of Information Science, 1945–85*. New York: Academic Press, 1989.

Lyons, H. *The Royal Society*. London: Cambridge University Press, 1944.

MacBride, S. *Many Voices, One World*. London: Kogan Press, 1980.

Mantoux, P. *The Industrial Revolution in the Eighteenth Century*. London: Cage, 1961.

Manzer, B.M. *The Abstract Journal, 1790–1920*. Metuchen, N.J.: Scarecrow Press, 1977.

Marker, G. *Publishing, Printing and the Origins of Intellectual Life in Russia, 1700–1800*. Princeton: Princeton University Press, 1985.

Marks, J. *Science and the Making of the Modern World*. London: Heinemann, 1983.

Mason, S.F. *A History of the Sciences*. London: Routledge, 1953.

Mathews, M.M. *Survey of English Dictionaries*. New York: Russell, 1966.

McArthur, T. *Worlds of Reference*. London: Cambridge University Press, 1986.

McClellan, J.E. *Science Reorganized: Scientific Societies in the Eighteenth Century*. New York: Columbia University Press, 1985.

McHale, J. *The Changing Information Environment*. London: Elek, 1976.

McKie, D., ed. *Philosophical Magazine*, Commemoration Number 1948, 122–32.

McMurtrie, D.C. *The Book*. London: Oxford University Press, 1943.

Meadows, A.J. *Communication in Science*. London: Butterworths, 1974.

———, ed. *The Origins of Information Science*. vol.1, London: Taylor Graham, 1987.

Menabrea, L.F. *Sketch of the Analytical Engine Invented by Charles Babbage, with Notes upon the Memoir by the Translator Ada Lovelace*. Geneva: Bibliothèque Universelle de Genéve, 1843.

Merton, R.K. *Social Theory and Social Structure*. New York: Free Press, 1968.

Merz, J.T. *A History of Scientific Thought in the Nineteenth Century*. New York: Dover, 1965.

Mieli, Aldo. *La science arabe*. Leyden: Brill, 1939.

Mikhailov, A.I. et al. *Scientific Communication and Informatics*. Arlington, Va.: Information Resources Press, 1984.

Moles, A. *Communication et les mass media*. Paris: Marabout, 1973.

Moran, B.T. "Privilege, Communication and Chemistry." *Ambix* 32 (1985): 110–26.

Moravcsik, M.J. *Science Development: The Building of Science in Less Developed Countries*. Bloomington: Indiana University Press, 1975.

Mountstephens, B. et al. *Quantitative Data in Science and Technology*. London: Aslib, 1971.

Mumby, F.A. *Publishing and Bookselling*. London: Cape, 1934.

Mumford, Lewis. *The City in History*. London: Pelican, 1966.

———. *Technics and Civilisation*. London: Routledge, 1934.

Needham, J. *Clerks and Craftsmen in China and the West*. London: Cambridge University Press, 1970.

Nelson, C.E. and D.K. Pollock, ed. *Communication among Scientists and Engineers*. Lexington, Mass.: Heath, 1970.

Nordenskiold, E. *The History of Biology*. New York: Knopf, 1928.

O'Leary, De L. *How Greek Science Passed to the Arabs*. London: Routledge, 1949.

Olby, R., ed. *Companion to the History of Modern Science*. London: Routledge, 1997.

Ornstein, M. *Role of the Scientific Societies in the Seventeenth Century*. New York: Arno Press, 1975.

Oseman, R. *Conferences and Their Literature*. London: Library Association, 1989.

Overhage, C.F.J. and R.J. Harman. *Intrex*. Cambridge, Mass.: MIT Press, 1965.

Pacey, A. *Technology in World Civilisation*. Oxford: Blackwell, 1990.

Parsons, E.A. *The Alexandrian Library*. London: Cleaver-Hume, 1952.

Passman, S. *Scientific and Technological Communication*. London: Pergamon, 1969.

Pickard, M.E. "Government and Science in the U.S.: Historical Backgrounds." *Journal of History of Medicine* 1 (1946): 254–89 and 446–81.

Pinner, H.L. *The World of Books in Classical Antiquity*. Leyden: Sijthoff, 1958.

Posner, E. *Archives in the Ancient World*. Cambridge, Mass.: Harvard University Press, 1972.

Price, D.J. de Solla. "Is Technology Historically Independent of Science?" *Technology and Culture* 6 (1965): 553–68.

———. *Little Science, Big Science*. New York: Columbia University Press, 1963.

———. "Measuring the Size of Science." *Procedures of the Israel Academy of Sciences* 4 (1969): 1–23.

———. "Networks of Scientific Papers." *Science* 149 (1965): 510–15.

———. *Science since Babylon*. New Haven: Yale University Press, 1961.

———. "The Structures of Publication in Science and Technology." In Gruber and Marquis, eds., *Factors in the Transfer of Technology*. Cambridge, Mass.: MIT Press, 1969.

Purver, M. *The Royal Society: Concept and Creation*. London: Routledge, 1967.

Putnam, G.H. *Authors and Their Public in Ancient Times*. New York: Putnam, 1894.

———. *Books and Their Makers during the Middle Ages*. New York: Putnam, 1896.

Pyenson, L. and Sheets-Pyenson, S. *Science in Society*. London: Fontana, 1998.

Quarterman, J.S. *The Matrix: Computer Networks and Conferencing Systems Worldwide*. Bedford, Mass.: Digital Press, 1990.

Raven, C.E. *English Naturalists from Neckam to Ray*. London: Cambridge University Press, 1947.

Ravetz, J.R. *Scientific Knowledge and Its Social Problems*. Oxford: Clarendon Press, 1971.

Rayleigh, Lord. Report of the 54th Meeting of the British Association for the Advancement of Science. London, 1884.

Reynolds, L.D. and N.G. Wilson. *Scribes and Scholars*. Oxford: Clarendon Press, 1974.

Roberts, J.M. *History of the World*. London: Hutchinson, 1976.

Rogers, E.M. and F.F. Shoemaker. *Communication of Innovations*. New York: Free Press, 1971.

Rose, H. and S. Rose. *Science and Society*. Harmondsworth: Penguin, 1969.

Rosenbloom, R.S. and F.W. Wolek. *Technology, Information and Organization: Information Transfer in Industrial R and D*. Cambridge, Mass.: Harvard University, Graduate School of Business Administration, 1967.

Royal Society Scientific Information Conference. *Proceedings*. London: Royal Society, 1948.

Russell, C.A. *Science and Social Change, 1700–1900*. London: MacMillan, 1983.

Samaran, C., ed. *L'histoire et ses methodes*. Paris: Gallimard, 1961.

Sandys, J.E. *History of Classical Scholarship*. London: Cambridge University Press, 1903–8.

Sansom, G.B. *The Western World and Japan*. London: Cresset Press, 1951.

Sarton, G. *Ancient Science and Modern Civilization*. Lincoln: University of Nebraska Press, 1954.

———. *Horus: A Guide to the History of Science*. New York: Ronald Press, 1952.

———. *Introduction to the History of Science*. Baltimore: Williams and Wilkins, 1927–48.

———. "The Scientific Literature Transmitted through the Incunabula." *Osiris* 5 (1938): 41–245.

Savory, T.H. *The Language of Science*. London: Deutsch, 1967.

Scharf, B. *Engineering and Its Language*. London: Muller, 1971.

Scudder, S.H. *Catalogue of Scientific Serials, 1633–1876*. Cambridge, Mass.: Harvard University Press, 1879.

Serres, M., ed. *History of Scientific Thought.* Oxford: Blackwell, 1995.
Shera, J.H. "History and Foundations of Information Science." *Annual Review of Information Science and Technology* 12 (1977): 249–75.
Singer, C. *Short History of Scientific Ideas.* Oxford: Clarendon Press, 1959.
Singer, C. et al. *The History of Technology.* London: Oxford University Press, 1954–8.
Small, H.G. and B.C. Griffith. "The Structure of Scientific Literature." *Science Studies* 4 (1974): 17–40 and 339–65.
Stearn, W.T. *Botanical Latin.* Newton Abbot: David and Charles, 1973.
Steinberg, S.H. *Five Hundred Years of Printing.* Harmondsworth: Penguin, 1974.
Stimson, D. *Scientists and Amateurs: A History of the Royal Society.* London: Sigma Books, 1949.
Storer, N.W. *The Social System of Science.* New York: Holt, 1966.
Taton, R., ed. *General History of the Sciences.* London: Thames and Hudson, 1963–66.
Thieme, P. *Scientific American,* October 1958.
Thomas, B.J. *The Internet for Scientists and Engineers.* Oxford: Oxford University Press, 1997.
Thompson, J.W. *The Medieval Library.* New York: Hafner, 1957.
Thomson, M.J. *Trade Literature: A Review and a Survey.* London: Science Reference Library, 1977.
Thornton, J.L. *Medical Books, Libraries and Collectors.* London: Deutsch, 1966.
Thornton, J.L. and R.I.J. Tully. *Scientific Books, Libraries and Collectors.* London: Library Association, 1971.
Toynbee, A. *A Study of History.* Abridged edition, London: Oxford University Press, 1972.
Turner, G.L., ed. *The Patronage of Science in the Nineteenth Century.* Leyden: Noordhoff, 1976.
Turnbull, G.H. *Hartlib, Dury and Comenius.* London: Liverpool University Press, 1947.
UNISIST: Study Report on the Feasibility of a World Science Information System. Paris: UNESCO, 1971.
United Nations Development Programme. *Human Development Report 1992.* Oxford: Oxford University Press, 1992.
Van Hoesen, H.B. and F.K. Walter. *Bibliography.* New York: Scribner, 1928.
Vickery, B.C. "Science, Research and History." *Stechert-Hefner Book News* 13 (1959): 65–6.
———. "Scientific Information: Problems and Prospects." *Minerva* 2 (1963): 21–38.
———. "The Growth of Scientific Literature, 1660–1970." In D.J. Foskett, ed. *The Information Environment.* Amsterdam: Elsevier, 1990.
Vickery, B.C. and A. Vickery. *Information Science in Theory and Practice.* London: Bowker-Saur, 1992.

Walzer, R. "Arabic Transmision of Greek Thought to Medieval Europe." *John Rylands Library Bulletin* 29 (1945): 1–26.

Webster, C. *The Great Instauration*. London: Duckworth, 1975.

Weinberg, A.M. *Reflections on Big Science*. Cambridge, Mass.: MIT Press, 1967.

Weinberg, A.M. et al. *Science, Government and Information*. Cambridge, Mass.: MIT Press, 1963.

Weiss, P. "Knowledge: A Growth Process." *Science* 131 (1960): 1716–19.

Weissgerber, D.W. "Interdisciplinary Searching." *Journal of Documentation* 49 (1993): 231–54.

Wells, H.G. *The Outline of History*. London: Cassell, 1951.

Whewell, W. *History of the Inductive Sciences*. London: Parker, 1857.

———. *Novum Organon Renovatum*. London: Parker, 1858.

White, L. *Mediaeval Technology and Social Change*. London: Oxford University Press, 1962.

Williams, R., ed. *Contact: Human Communication and Its History*. London: Thames and Hudson, 1981.

Winter, H.J.J. *Eastern Science*. London: Murray, 1952.

Wolf, A. *History of Science, Technology and Philosophy in the Sixteenth and Seventeenth Centuries*. London: Allen and Unwin, 1951.

———. *History of Science, Technology and Philosophy in the Eighteenth Century*. London: Allen and Unwin, 1952.

Wolff, P. *The Awakening of Europe*. Harmondsworth: Penguin, 1968.

———. *Western Languages, AD 100–1500*. London: Weidenfeld, 1971.

Wood, D.N. "The Foreign Language Problem." *Journal of Documentation* 23 (1967): 117–30.

Woods, R.S. *The Naturalist's Lexicon*. Pasadena: Abbey Garden Press, 1944.

Woodward, A.M. *Problems and Possible Investigations in the Study of the Role of Reviews in Information Transfer in Science*. London: Aslib, 1975.

Woodward, C.D. *BSI: The Story of Standards*. London: British Standards Institution, 1972.

Ziman, J. *Public Knowledge*. London: Cambridge University Press, 1968.

———. *The Force of Knowledge*. London: Cambridge University Press, 1976.

Index

About the Author

Brian C. Vickery is Professor Emeritus in Librarianship at University College London. He received an M.A. in chemistry from the University of Oxford in 1941. He worked at a chemical munitions factory during 1941–45, then joined Imperial Chemical Industries as an information officer. In 1960 he became Principal Scientific Officer at the U.K. National Library for Science and Technology. After a spell as Librarian at the University of Manchester Institute of Science and Technology, he was during 1966–73 the Research Director of Aslib, the U.K. institute for information management. From 1973 to 1983 he was Professor and Director of the School of Library and Information Studies at University College London, and since then has worked as a private consultant on information matters.

His previous books include *Classification and Indexing in Science, On Retrieval System Theory, Information Systems*, and *Information Science in Theory and Practice.* He has undertaken consultancies for the International Council of Scientific Unions, the UN Food and Agriculture Organization, the Commonwealth Agricultural Bureaux, the Institution of Mechanical Engineers, the European Commission, and UNESCO. His abiding interests have been the transmission of scientific information, the application of computers and telecommunications to this, and the history and development of science and technology.